INTERMEDIATE 2

PHYSICS
2008-2012

First exam published in 2008.
Published by Bright Red Publishing Ltd, 6 Stafford Street, Edinburgh EH3 7AU
tel: 0131 220 5804 fax: 0131 220 6710 info@brightredpublishing.co.uk www.brightredpublishing.co.uk

ISBN 978-1-84948-279-0

A CIP Catalogue record for this book is available from the British Library.

Bright Red Publishing is grateful to the copyright holders, as credited on the final page of the Question Section, for permission to use their material. Every effort has been made to trace the copyright holders and to obtain their permission for the use of copyright material. Bright Red Publishing will be happy to receive information allowing us to rectify any error or omission in future editions.

INTERMEDIATE 2

2008

[BLANK PAGE]

X069/201

NATIONAL QUALIFICATIONS 2008

FRIDAY, 23 MAY
1.00 PM – 3.00 PM

PHYSICS
INTERMEDIATE 2

Read Carefully

Reference may be made to the Physics Data Booklet

1 All questions should be attempted.

Section A (questions 1 to 20)

2 Check that the answer sheet is for Physics Intermediate 2 (Section A).

3 For this section of the examination you must use an **HB pencil** and, where necessary, an eraser.

4 Check that the answer sheet you have been given has **your name**, **date of birth**, **SCN** (Scottish Candidate Number) and **Centre Name** printed on it.

Do not change any of these details.

5 If any of this information is wrong, tell the Invigilator immediately.

6 If this information is correct, **print** your name and seat number in the boxes provided.

7 There is **only one correct** answer to each question.

8 Any rough working should be done on the question paper or the rough working sheet, **not** on your answer sheet.

9 At the end of the exam, put the **answer sheet for Section A inside the front cover of your answer book**.

10 Instructions as to how to record your answers to questions 1–20 are given on page three.

Section B (questions 21 to 31)

11 Answer the questions numbered 21 to 31 in the answer book provided.

12 **All answers must be written clearly and legibly in ink.**

13 Fill in the details on the front of the answer book.

14 Enter the question number clearly in the margin of the answer book beside each of your answers to questions 21 to 31.

15 Care should be taken to give an appropriate number of significant figures in the final answers to calculations.

LI X069/201 6/7870

DATA SHEET

Speed of light in materials

Material	*Speed* in m/s
Air	$3{\cdot}0 \times 10^8$
Carbon dioxide	$3{\cdot}0 \times 10^8$
Diamond	$1{\cdot}2 \times 10^8$
Glass	$2{\cdot}0 \times 10^8$
Glycerol	$2{\cdot}1 \times 10^8$
Water	$2{\cdot}3 \times 10^8$

Speed of sound in materials

Material	*Speed* in m/s
Aluminium	5200
Air	340
Bone	4100
Carbon dioxide	270
Glycerol	1900
Muscle	1600
Steel	5200
Tissue	1500
Water	1500

Gravitational field strengths

	Gravitational field strength on the surface in N/kg
Earth	10
Jupiter	26
Mars	4
Mercury	4
Moon	1·6
Neptune	12
Saturn	11
Sun	270
Venus	9

Specific heat capacity of materials

Material	*Specific heat capacity* in J/kg °C
Alcohol	2350
Aluminium	902
Copper	386
Glass	500
Ice	2100
Iron	480
Lead	128
Oil	2130
Water	4180

Specific latent heat of fusion of materials

Material	*Specific latent heat of fusion* in J/kg
Alcohol	$0{\cdot}99 \times 10^5$
Aluminium	$3{\cdot}95 \times 10^5$
Carbon dioxide	$1{\cdot}80 \times 10^5$
Copper	$2{\cdot}05 \times 10^5$
Iron	$2{\cdot}67 \times 10^5$
Lead	$0{\cdot}25 \times 10^5$
Water	$3{\cdot}34 \times 10^5$

Melting and boiling points of materials

Material	*Melting point* in °C	*Boiling point* in °C
Alcohol	−98	65
Aluminium	660	2470
Copper	1077	2567
Glycerol	18	290
Lead	328	1737
Iron	1537	2747

Specific latent heat of vaporisation of materials

Material	*Specific latent heat of vaporisation* in J/kg
Alcohol	$11{\cdot}2 \times 10^5$
Carbon dioxide	$3{\cdot}77 \times 10^5$
Glycerol	$8{\cdot}30 \times 10^5$
Turpentine	$2{\cdot}90 \times 10^5$
Water	$22{\cdot}6 \times 10^5$

Radiation weighting factors

Type of radiation	*Radiation weighting factor*
alpha	20
beta	1
fast neutrons	10
gamma	1
slow neutrons	3

SECTION A

For questions 1 to 20 in this section of the paper the answer to each question is either A, B, C, D or E. Decide what your answer is, then, using your pencil, put a horizontal line in the space provided—see the example below.

EXAMPLE

The energy unit measured by the electricity meter in your home is the

A kilowatt-hour

B ampere

C watt

D coulomb

E volt.

The correct answer is **A**—kilowatt-hour. The answer **A** has been clearly marked in **pencil** with a horizontal line (see below).

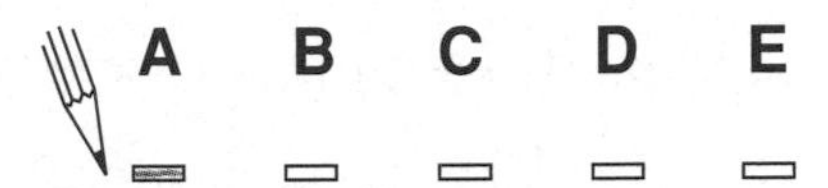

Changing an answer

If you decide to change your answer, carefully erase your first answer and, using your pencil, fill in the answer you want. The answer below has been changed to **E**.

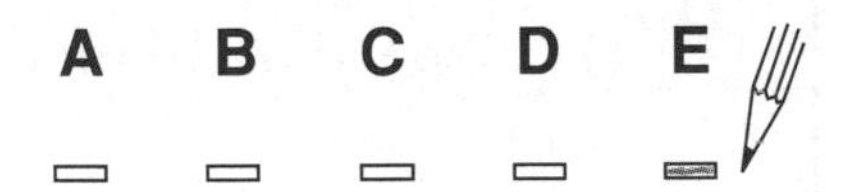

[Turn over

SECTION A

Answer questions 1–20 on the answer sheet.

1. Which of the following is a vector quantity?

A Distance

B Energy

C Speed

D Time

E Velocity

2. A student walks from X to Y and then from Y to Z.

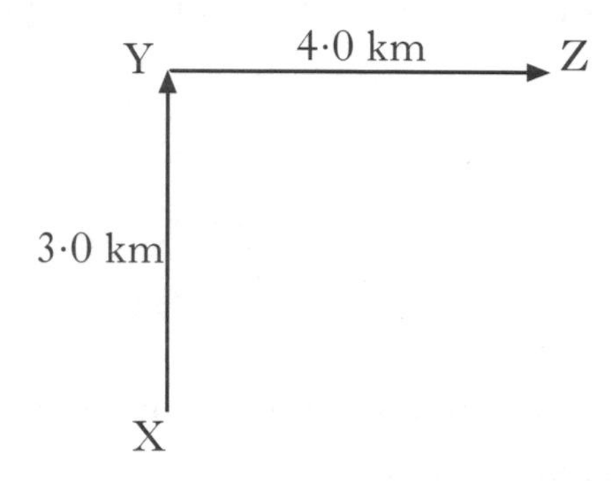

The complete walk takes 2 hours.

Which row in the table shows the average speed and the average velocity for the complete walk?

	Average speed	*Average velocity*
A	2·5 km/h	2·5 km/h at 053
B	2·5 km/h at 053	2·5 km/h
C	3·5 km/h	2·5 km/h at 053
D	3·5 km/h at 053	3·5 km/h
E	3·5 km/h	3·5 km/h at 053

3. A car travelling in a straight line decelerates uniformly from 20 m/s to 12 m/s in 4 seconds. The displacement of the car in this time is

A 32 m

B 48 m

C 64 m

D 80 m

E 128 m.

4. An unbalanced force of one newton will make a

A 0·1 kg mass accelerate at 1 m/s^2

B 1 kg mass accelerate at 1 m/s^2

C 1 kg mass accelerate at 10 m/s^2

D 0·1 kg mass move at a constant speed of 1 m/s

E 1 kg mass move at a constant speed of 10 m/s.

5. A trolley of mass 0·6 kg is travelling at 5 m/s along a smooth, level track.

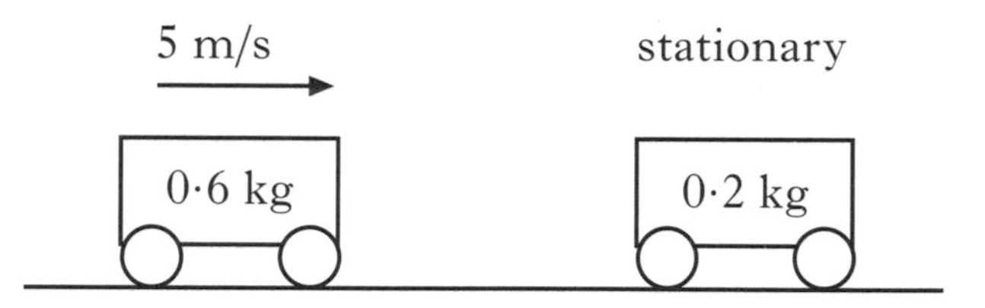

The trolley collides with a stationary trolley of mass 0·2 kg.

The magnitude of the total momentum of the trolleys immediately after collision is

A 0 kg m/s

B 1·0 kg m/s

C 2·0 kg m/s

D 3·0 kg m/s

E 4·0 kg m/s.

6. A power station has an efficiency of 40%. The input power to the station is 1600 MW.

What is the useful output power?

A 40 MW

B 640 MW

C 960 MW

D 4000 MW

E 64000 MW

7. A sample of water is at a temperature of 100 °C. The sample absorbs $2·3 \times 10^4$ J of energy.

The specific latent heat of vaporisation of water is $22·6 \times 10^5$ J/kg.

The mass of water changed into steam at 100 °C is

A 0·01 kg

B 5·3 kg

C 100 kg

D $2·3 \times 10^4$ kg

E $2·3 \times 10^6$ kg.

8. Three circuit symbols X, Y and Z are shown.

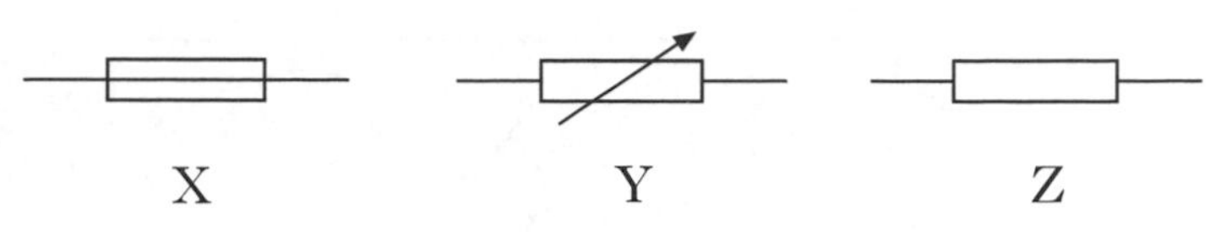

Which row in the table identifies the symbols X, Y and Z?

	X	Y	Z
A	thermistor	transistor	resistor
B	fuse	variable resistor	thermistor
C	transistor	fuse	variable resistor
D	fuse	variable resistor	resistor
E	variable resistor	resistor	fuse

9. A circuit is set up as shown.

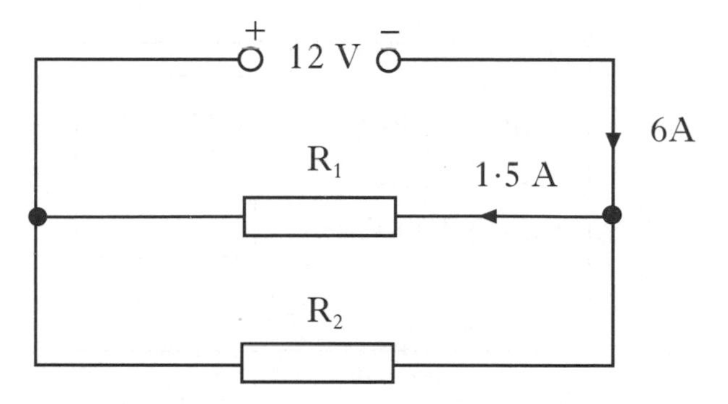

The current from the supply is 6 A. The current in resistor R_1 is 1·5 A.

Which row in the table shows the potential difference across resistor R_2 and the current in resistor R_2?

	Potential difference across R_2 (V)	*Current in R_2* (A)
A	12	1·5
B	6	1·5
C	12	4·5
D	6	4·5
E	12	7·5

[Turn over

10. A circuit is set up as shown.

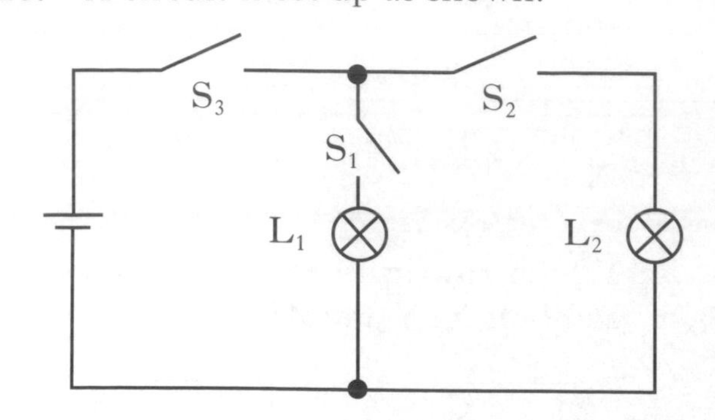

Which switch or switches must be closed to light lamp L_1 **only**?

A S_1 only

B S_2 only

C S_1 and S_2 only

D S_1 and S_3 only

E S_2 and S_3 only

11. The information shown is for an electric food mixer.

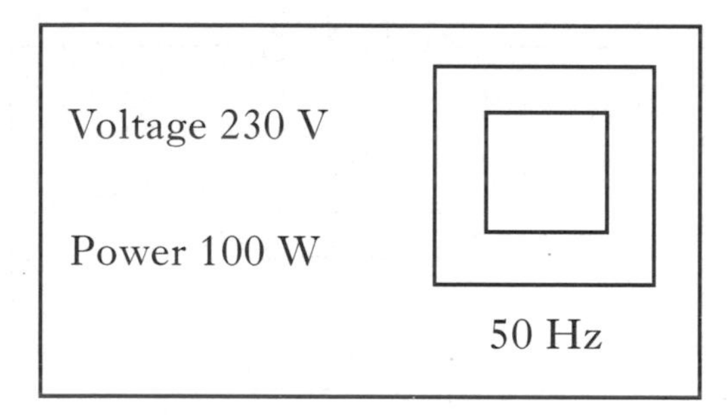

The resistance of the mixer is

A 0·43 Ω

B 2·3 Ω

C 4·6 Ω

D 529 Ω

E 23 000 Ω.

12. When a magnet is pushed into or pulled out of a coil of wire, a voltage is induced across the ends of the coil.

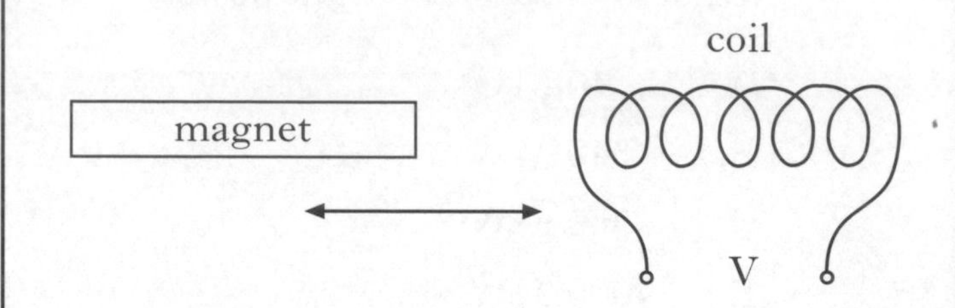

Which of the following produces the greatest induced voltage?

	Strength of magnet	*Speed of magnet*	*Number of turns in a coil*
A	weak	slow	20
B	weak	fast	40
C	strong	slow	20
D	strong	fast	20
E	strong	fast	40

13. A manufacturer states that an amplifier has a voltage gain of 15. This means that

A the output frequency is 15 times the input frequency

B the input frequency is 15 times the output frequency

C the output voltage is 15 times the input voltage

D the input voltage is 15 times the output voltage

E the input voltage is 15 V.

14. The following diagram shows a wave.

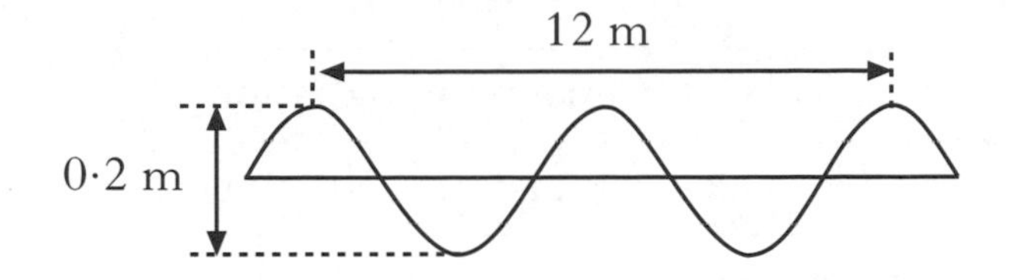

Which row in the table gives the wavelength and amplitude of the wave?

	Wavelength (m)	*Amplitude* (m)
A	4	0·2
B	6	0·1
C	6	0·2
D	12	0·1
E	12	0·2

15. A ray of light passes from air into a glass block as shown.

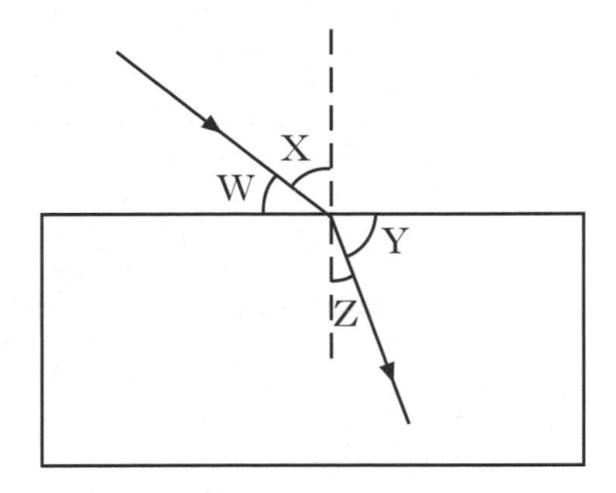

Which row in the table shows the angle of incidence and the angle of refraction?

	Angle of incidence	*Angle of refraction*
A	W	Z
B	W	Y
C	X	Z
D	X	Y
E	Z	X

16. A student wears glasses fitted with concave lenses. Which of the following statements is/are correct?

I The student is short sighted.

II Concave lenses are converging lenses.

III The glasses help the student to see near objects clearly.

A I only

B II only

C III only

D I and II only

E I, II and III

[Turn over

17. Which row in the table describes an alpha particle, a beta particle and a gamma ray?

	Alpha particle	*Beta particle*	*Gamma ray*
A	neutron	helium nucleus	electromagnetic radiation
B	helium nucleus	electron	electromagnetic radiation
C	hydrogen nucleus	electromagnetic radiation	electron
D	helium nucleus	electromagnetic radiation	neutron
E	hydrogen nucleus	electron	electromagnetic radiation

18. For a particular radioactive source, 1800 atoms decay in a time of 3 minutes. The **activity** of this source is

A 10 Bq

B 600 Bq

C 1800 Bq

D 5400 Bq

E 324 000 Bq.

19. One gray is equal to

A one becquerel per kilogram

B one sievert per second

C one joule per second

D one sievert per kilogram

E one joule per kilogram.

20. A student makes the following statements about nuclear reactors.

I Fission takes place in the fuel rods.

II The material in the control rods slows down neutrons.

III The material in the moderator absorbs neutrons.

Which of the statements is/are correct?

A I only

B I and II only

C I and III only

D II and III only

E I, II and III

SECTION B

Marks

Write your answers to questions 21–31 in the answer book.

All answers must be written clearly and legibly in ink.

21. Athletes in a race are recorded by a TV camera which runs on rails beside the track.

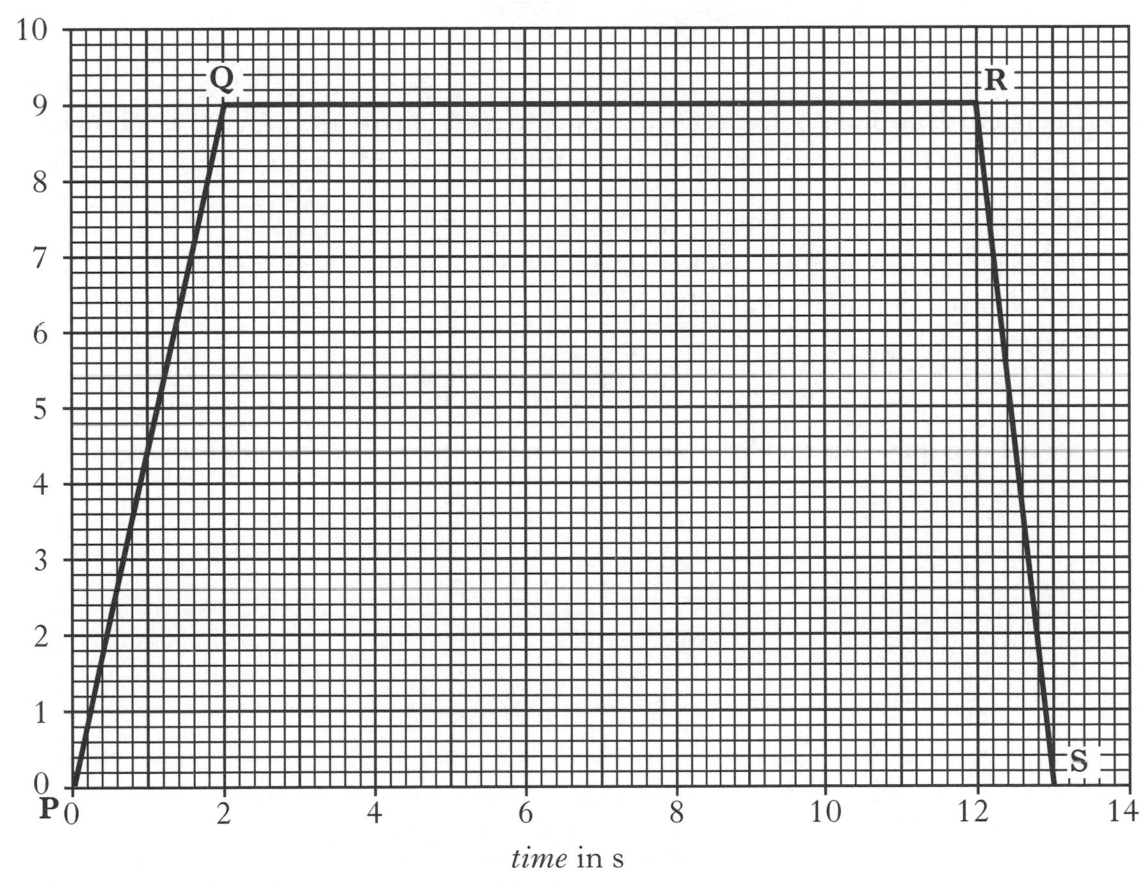

The graph shows the speed of the camera during the race.

(*a*) Calculate the acceleration of the camera between **P** and **Q**. **2**

(*b*) The mass of the camera is 15 kg.

Calculate the unbalanced force needed to produce the acceleration between **P** and **Q**. **2**

(*c*) How far does the camera travel in the 13 s? **2**

(*d*) The camera lens has a focal length of 200 mm.

Calculate the power of the lens. **2**

(8)

Marks

22. A fairground ride uses a giant catapult to launch people upwards using elastic cords.

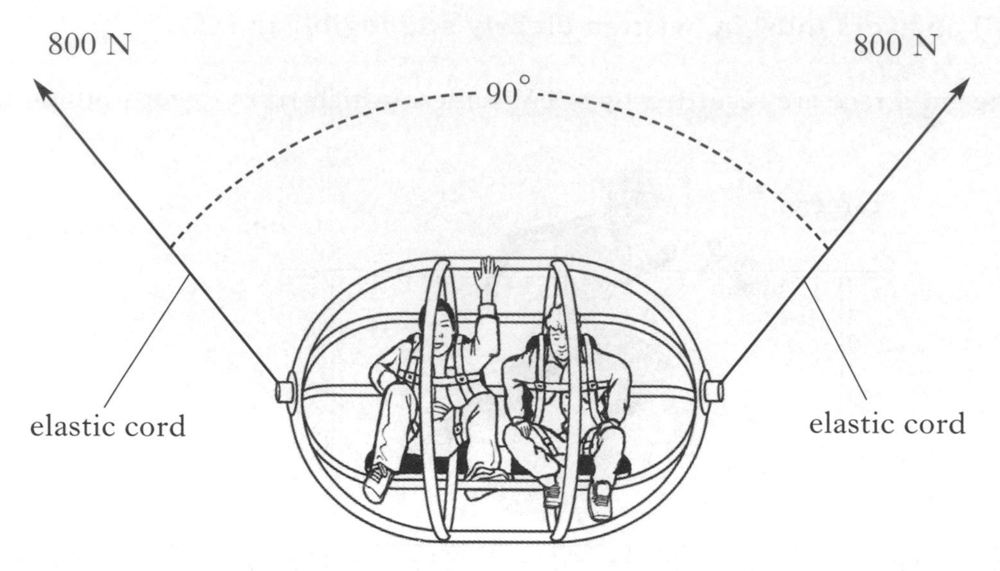

(*a*) Each cord applies a force of 800 N and the cords are at 90° as shown. Using a scale diagram, or otherwise, find the size of the resultant of these two forces. **2**

(*b*) The cage is now pulled further down before release. The cords provide an upward resultant force of 2700 N. The cage and its occupants have a total mass of 180 kg.

(i) Calculate the weight of the cage and occupants. **2**

(ii) Calculate the acceleration of the cage and occupants when released. **3**

(7)

Marks

23. One type of exercise machine is shown below.

(*a*) A person using this machine pedals against friction forces applied to the wheel by the brake.

A friction force of 300 N is applied at the edge of the wheel, which has a circumference of 1·5 m.

(i) How much work is done by friction in one turn of the wheel? **2**

(ii) The person turns the wheel 500 times in 5 minutes.

Calculate the average power produced. **3**

(*b*) The wheel is a solid aluminium disc of mass 12·0 kg.

(i) All the work done by friction is converted to heat in the disc.

Calculate the temperature rise after 500 turns. **2**

(ii) Explain why the actual temperature rise of the disc is less than calculated in (*b*) (i). **1**

(8)

[Turn over

Marks

24. An early method of crash testing involved a car rolling down a slope and colliding with a wall.

In one test, a car of mass 750 kg starts at the top of a 7·2 m high slope.

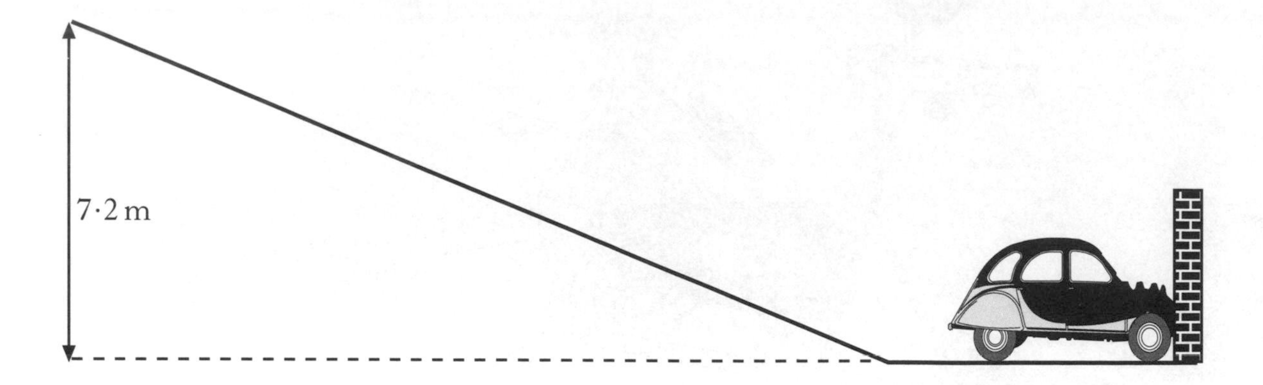

(*a*) Calculate the gravitional potential energy of the car at the top of the slope. **2**

(*b*) (i) State the value of the kinetic energy of the car at the bottom of the slope, assuming no energy losses. **1**

(ii) Calculate the speed of the car at the bottom of the slope, before hitting the wall. **2**

(5)

Marks

25. Some resistors are labelled with a power rating as well as their resistance value. This is the maximum power at which they can operate without overheating.

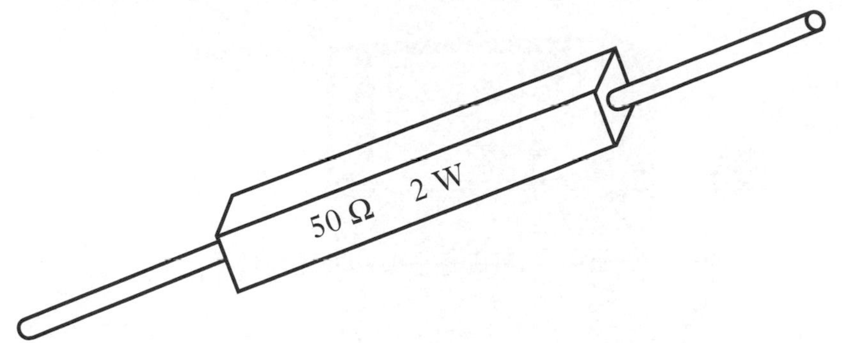

(*a*) A resistor is labelled 50 Ω, 2 W.

Calculate the maximum operating current for this resistor. **2**

(*b*) Two resistors, each rated at 2 W, are connected in parallel to a 9 V d.c. supply.

They have resistances of 60 Ω and 30 Ω.

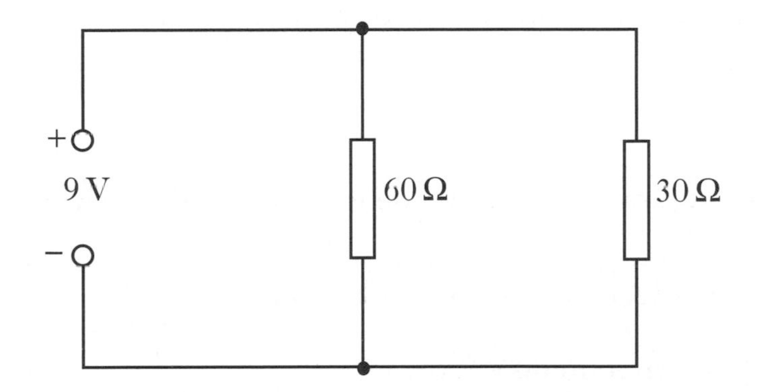

(i) Calculate the total resistance of the circuit. **2**

(ii) Calculate the power produced in each resistor. **3**

(iii) State which, if any, of the resistors will overheat. **1**

(*c*) The 9 V **d.c.** supply is replaced by a 9 V **a.c.** supply.

What effect, if any, would this have on your answers to part (*b*) (ii)? **1**

(9)

[Turn over

Marks

26. A karaoke machine consists of a microphone, amplifier, loudspeaker, DVD player and screen.

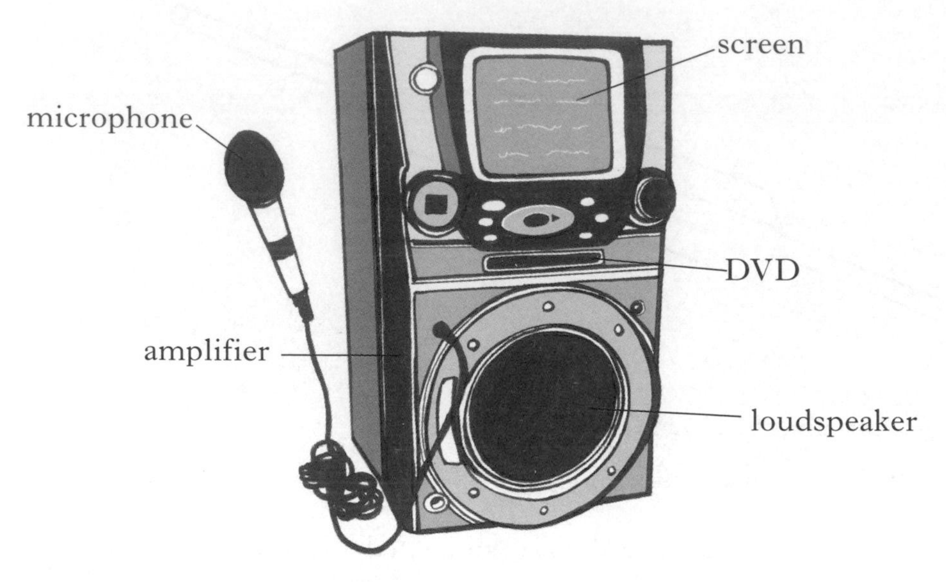

(*a*) What energy change takes place in the microphone? **1**

(*b*) The amplifier processes the signal from the microphone.

What effect does the amplifier have on the signal's

(i) frequency; **1**

(ii) amplitude? **1**

(*c*) A singer produces a note of frequency 850 Hz. The speed of sound in air is 340 m/s.

Calculate the wavelength of this note in air. **2**

(*d*) The DVD player contains a laser.

Light from this laser enters a small glass prism as shown.

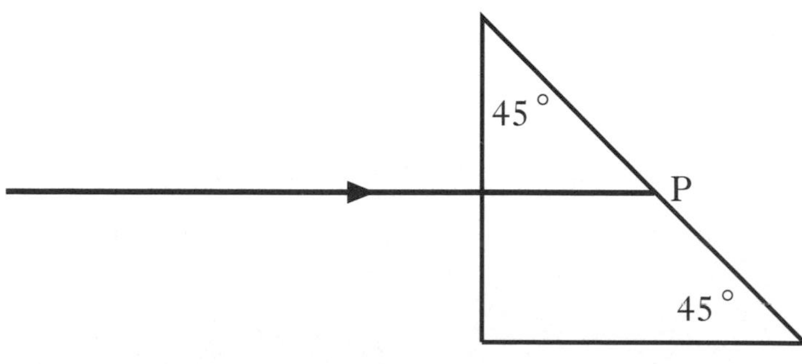

The glass has a critical angle of 40 °.

(i) Explain what is meant by the term "critical angle". **1**

(ii) Copy and complete the diagram to show the path of the ray after it strikes point P. **2**

(8)

Marks

27. An office has an automatic window blind that closes when the light level outside gets too high.

The electronic circuit that operates the motor to close the blind is shown.

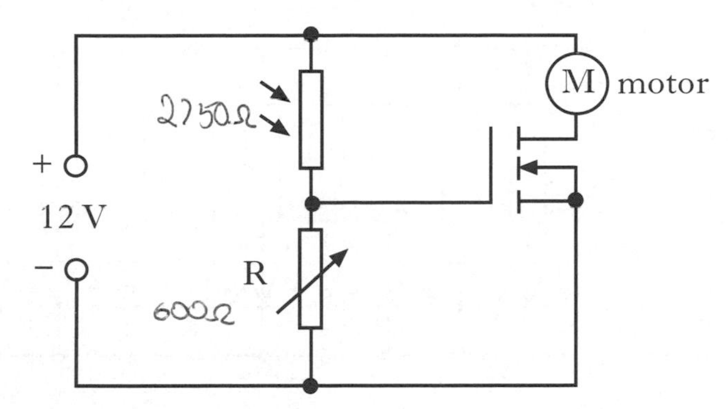

(*a*) The MOSFET switches on when the voltage across variable resistor R reaches 2·4 V.

(i) Explain how this circuit works to close the blind. **3**

(ii) What is the purpose of the variable resistor R? **1**

(*b*) The graph shows how the resistance of the LDR varies with light level.

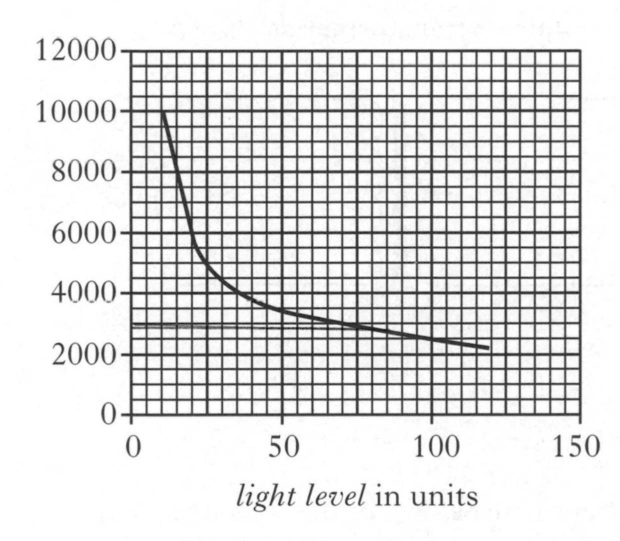

(i) What is the resistance of the LDR when the light level is 70 units? **1**

(ii) R has a value of 600 Ω. Calculate the voltage across R when the light level is 70 units. **2**

(iii) State whether or not the blinds will close when the light level is 70 units.

Justify your answer. **2**

(9)

[Turn over

Marks

28. The rear light of a car is made up of a row of 10 **identical** red LEDs. Each LED requires 2 V and 20 mA to operate correctly.

(*a*) The circuit for this is shown.

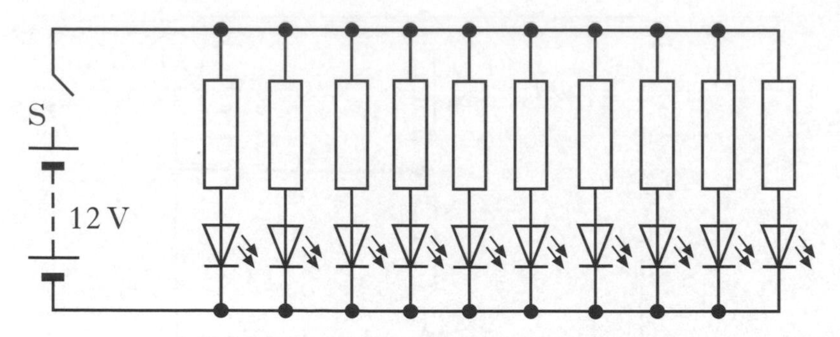

(i) Why does each LED need a resistor in series? **1**

(ii) The voltage of the car battery is 12 V.

Calculate the value of each resistor. **3**

(iii) Calculate the total current, **in amperes**, from the battery when the rear light is operating correctly. **2**

(*b*) Some car headlights require 84 V to operate. Electronic circuits are needed to convert the car battery voltage.

Part of the circuit contains a transformer as shown.

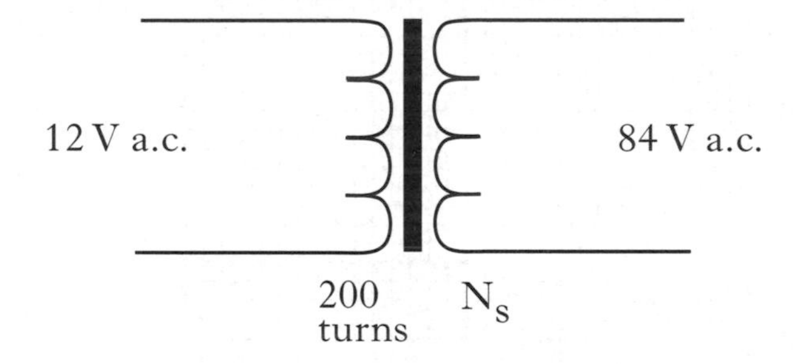

The primary coil of the transformer has 200 turns.

Calculate the number of turns, N_S, in the secondary coil. **2**

(8)

Marks

29. A "bug viewer" has a plastic chamber with a lens in the lid. It is used to get a magnified view of small insects placed on the base of the chamber.

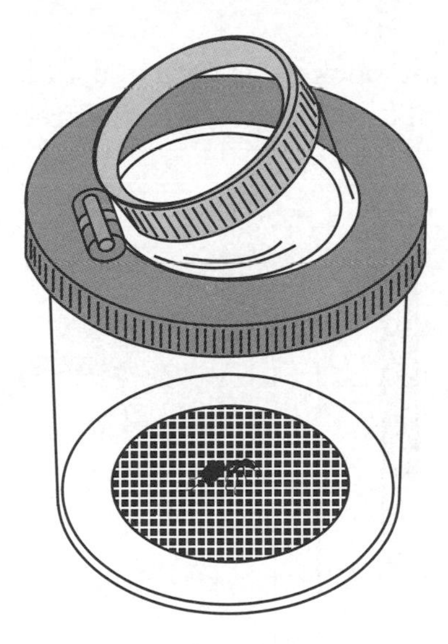

(*a*) What type of lens should be used? **1**

(*b*) The lens used has a focal length of 60 mm and the base of the chamber is 30 mm from the lens.

Copy and complete this diagram by adding rays to show where the image of the bug will be formed. **2**

Use the squared ruled paper provided (small squares side).

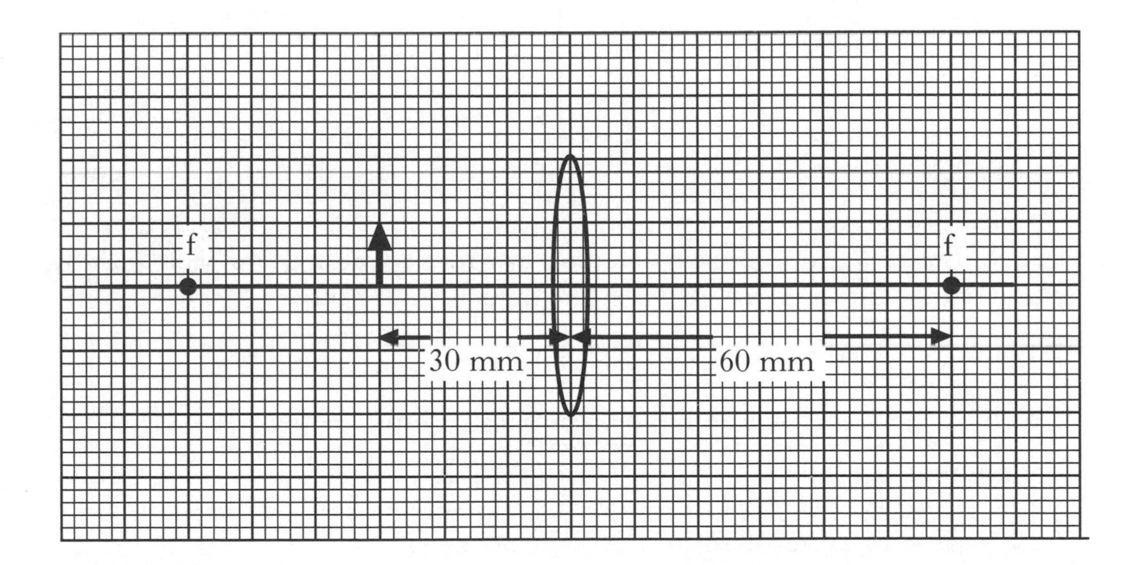

(*c*) How would the shape of this lens have to be altered to give it a longer focal length? **1**

(*d*) Name the eye defect which this type of lens could correct. **1**

(5)

[Turn over

Marks

30. When welders join thick steel plates it is important that the joint is completely filled with metal. This ensures there are no air pockets in the metal weld, as this would weaken the joint.

One method of checking for air pockets is to use a radioactive source on one side of the joint. A detector placed as shown measures the count rate on the other side.

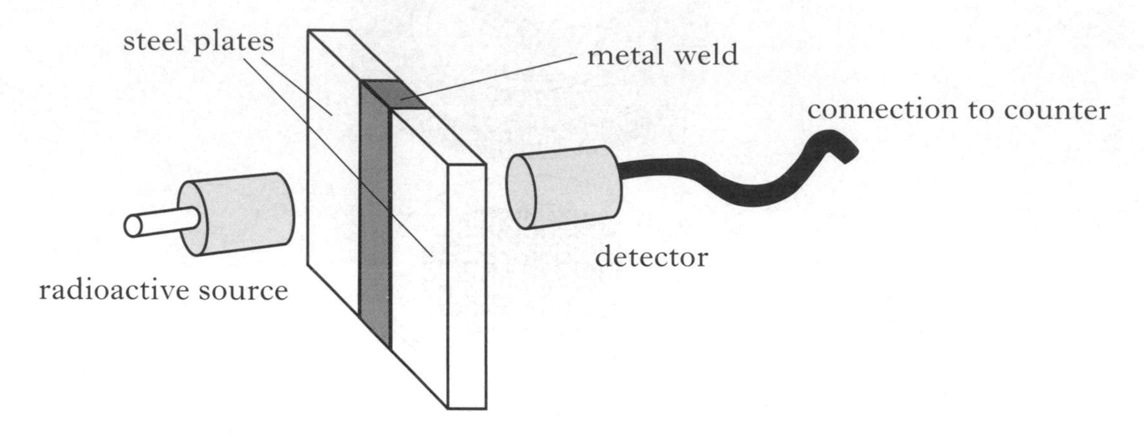

View from above

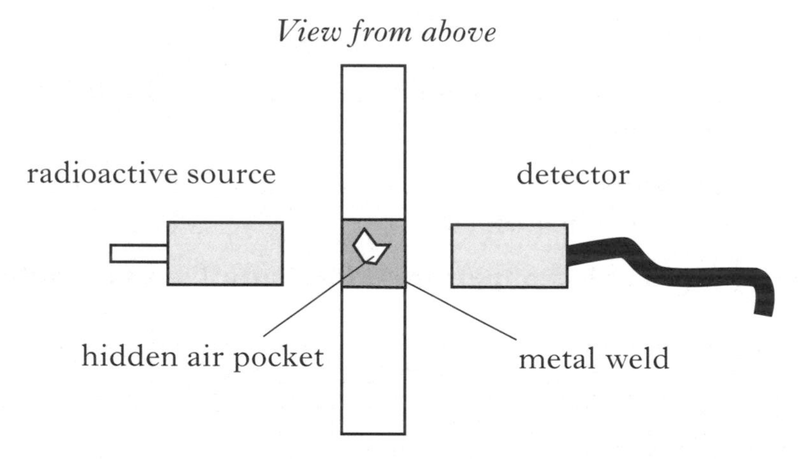

(*a*) The radioactive source and detector are moved along the weld. How would the count rate change when the detector moves over an air pocket?

Explain your answer. **2**

(*b*) Which of the radiations alpha, beta or gamma must be used?

Explain your answer. **2**

(*c*) X-rays are sometimes used to detect air pockets.

How does the wavelength of X-rays compare with gamma rays? **1**

(5)

Marks

31. Gold-198 is a radioactive source that is used to trace factory waste which may cause river pollution.

A small quantity of the radioactive gold is added into the waste as it enters the river. Scanning the river using radiation detectors allows scientists to trace where the waste has travelled.

Gold-198 has a half-life of 2·7 days.

(*a*) What is meant by the term "half-life"? **1**

(*b*) A sample of Gold-198 has an activity of 64 kBq when first obtained by the scientists.

Calculate the activity after 13·5 days. **2**

(*c*) Describe two precautions taken by the scientists to reduce the equivalent dose they receive while using radioactive sources. **2**

(*d*) A scientist receives an absorbed dose of 10 mGy of alpha radiation.

(i) Calculate the equivalent dose received. **2**

(ii) The risk of biological harm from radiation exposure depends on the absorbed dose and the type of radiation. Which other factor affects the risk of biological harm? **1**

(8)

[*END OF QUESTION PAPER*]

[BLANK PAGE]

INTERMEDIATE 2

2009

[BLANK PAGE]

X069/201

NATIONAL QUALIFICATIONS 2009

TUESDAY, 26 MAY
1.00 PM – 3.00 PM

PHYSICS
INTERMEDIATE 2

Read Carefully

Reference may be made to the Physics Data Booklet

1 All questions should be attempted.

Section A (questions 1 to 20)

2 Check that the answer sheet is for Physics Intermediate 2 (Section A).

3 For this section of the examination you must use an **HB pencil** and, where necessary, an eraser.

4 Check that the answer sheet you have been given has **your name**, **date of birth**, **SCN** (Scottish Candidate Number) and **Centre Name** printed on it.

Do not change any of these details.

5 If any of this information is wrong, tell the Invigilator immediately.

6 If this information is correct, **print** your name and seat number in the boxes provided.

7 There is **only one correct** answer to each question.

8 Any rough working should be done on the question paper or the rough working sheet, **not** on your answer sheet.

9 At the end of the exam, put the **answer sheet for Section A inside the front cover of your answer book**.

10 Instructions as to how to record your answers to questions 1–20 are given on page three.

Section B (questions 21 to 29)

11 Answer the questions numbered 21 to 29 in the answer book provided.

12 **All answers must be written clearly and legibly in ink**.

13 Fill in the details on the front of the answer book.

14 Enter the question number clearly in the margin of the answer book beside each of your answers to questions 21 to 29.

15 Care should be taken to give an appropriate number of significant figures in the final answers to calculations.

LI X069/201 6/7520

DATA SHEET

Speed of light in materials

Material	*Speed* in m/s
Air	$3\cdot0 \times 10^8$
Carbon dioxide	$3\cdot0 \times 10^8$
Diamond	$1\cdot2 \times 10^8$
Glass	$2\cdot0 \times 10^8$
Glycerol	$2\cdot1 \times 10^8$
Water	$2\cdot3 \times 10^8$

Speed of sound in materials

Material	*Speed* in m/s
Aluminium	5200
Air	340
Bone	4100
Carbon dioxide	270
Glycerol	1900
Muscle	1600
Steel	5200
Tissue	1500
Water	1500

Gravitational field strengths

	Gravitational field strength on the surface in N/kg
Earth	10
Jupiter	26
Mars	4
Mercury	4
Moon	1·6
Neptune	12
Saturn	11
Sun	270
Venus	9

Specific heat capacity of materials

Material	*Specific heat capacity* in J/kg °C
Alcohol	2350
Aluminium	902
Copper	386
Glass	500
Ice	2100
Iron	480
Lead	128
Oil	2130
Water	4180

Specific latent heat of fusion of materials

Material	*Specific latent heat of fusion* in J/kg
Alcohol	$0\cdot99 \times 10^5$
Aluminium	$3\cdot95 \times 10^5$
Carbon dioxide	$1\cdot80 \times 10^5$
Copper	$2\cdot05 \times 10^5$
Iron	$2\cdot67 \times 10^5$
Lead	$0\cdot25 \times 10^5$
Water	$3\cdot34 \times 10^5$

Melting and boiling points of materials

Material	*Melting point* in °C	*Boiling point* in °C
Alcohol	−98	65
Aluminium	660	2470
Copper	1077	2567
Glycerol	18	290
Lead	328	1737
Iron	1537	2747

Specific latent heat of vaporisation of materials

Material	*Specific latent heat of vaporisation* in J/kg
Alcohol	$11\cdot2 \times 10^5$
Carbon dioxide	$3\cdot77 \times 10^5$
Glycerol	$8\cdot30 \times 10^5$
Turpentine	$2\cdot90 \times 10^5$
Water	$22\cdot6 \times 10^5$

Radiation weighting factors

Type of radiation	*Radiation weighting factor*
alpha	20
beta	1
fast neutrons	10
gamma	1
slow neutrons	3

SECTION A

For questions 1 to 20 in this section of the paper the answer to each question is either A, B, C, D or E. Decide what your answer is, then, using your pencil, put a horizontal line in the space provided—see the example below.

EXAMPLE

The energy unit measured by the electricity meter in your home is the

A kilowatt-hour

B ampere

C watt

D coulomb

E volt.

The correct answer is **A—**kilowatt-hour. The answer **A** has been clearly marked in **pencil** with a horizontal line (see below).

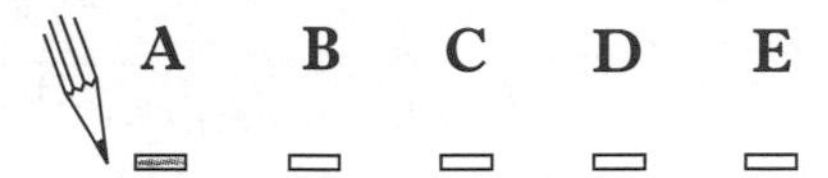

Changing an answer

If you decide to change your answer, carefully erase your first answer and, using your pencil, fill in the answer you want. The answer below has been changed to **E**.

A B C D E

[Turn over

SECTION A

Answer questions 1–20 on the answer sheet.

1. Which of the following quantities requires both magnitude and direction?

A Mass

B Distance

C Momentum

D Speed

E Time

2. A cross country runner travels 2·1 km North then 1·5 km East. The total time taken is 20 minutes.

The average speed of the runner is

A 0·18 m/s

B 2·2 m/s

C 3·0 m/s

D 130 m/s

E 180 m/s.

3. The graph shows how the velocity of an object varies with time.

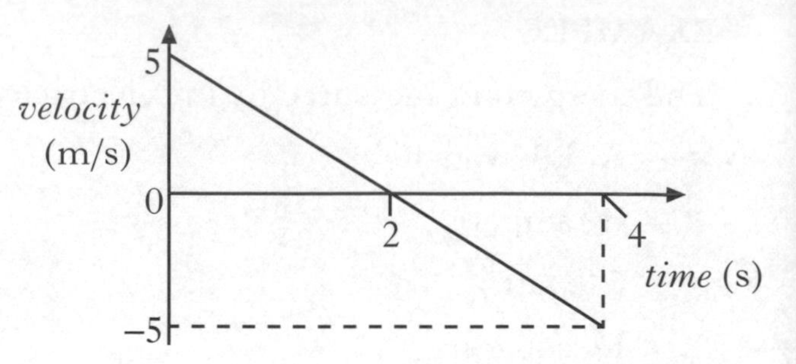

Which row in the table shows the displacement after 4 s and the acceleration of the object during the first 4 s?

	Displacement (m)	*Acceleration* (m/s^2)
A	10	–10
B	10	2·5
C	0	2·5
D	0	–10
E	0	–2·5

4 A ball is thrown horizontally from a cliff as shown.

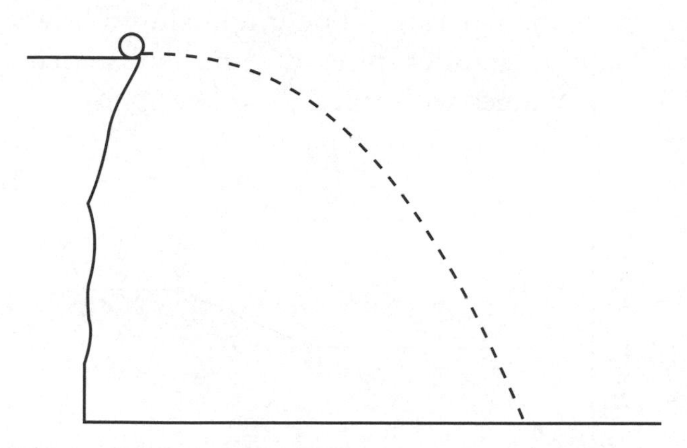

The effect of air resistance is negligible.

A student makes the following statements about the ball.

I The vertical speed of the ball increases as it falls.

II The vertical acceleration of the ball increases as it falls.

III The vertical force on the ball increases as it falls.

Which of the statements is/are correct?

A I only

B II only

C I and II only

D II and III only

E I, II and III

5. Which block has the largest resultant force acting on it?

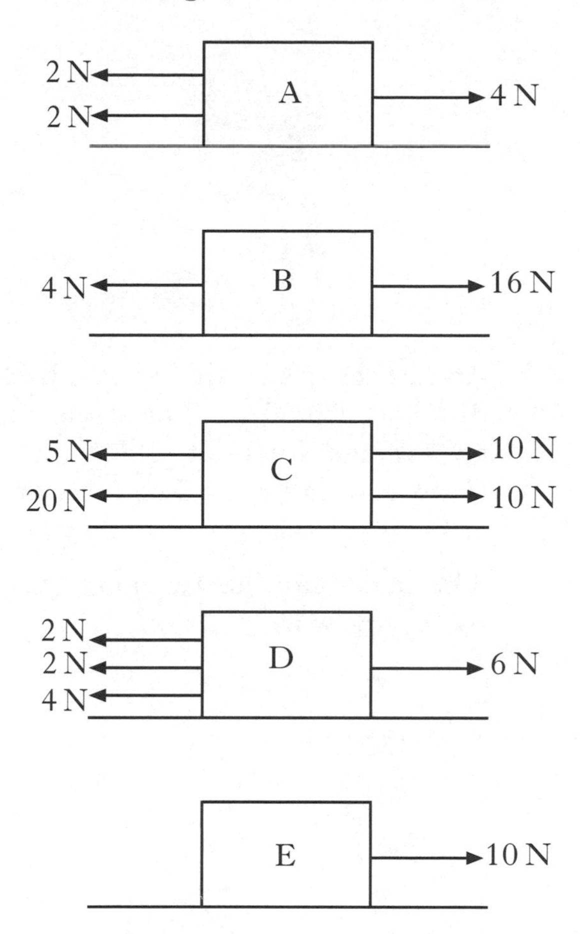

[Turn over

6. An arrow is fired from a bow as shown.

An archer pulls the string back a distance of 0·50 m. The string exerts an average force of 300 N on the arrow as it is fired. The mass of the arrow is 0·15 kg.

The maximum kinetic energy gained by the arrow is

A 23 J

B 150 J

C 600 J

D 2000 J

E 6750 J.

7. A solid substance is placed in an insulated container and is heated at a constant rate. The graph shows how the temperature of the substance changes with time.

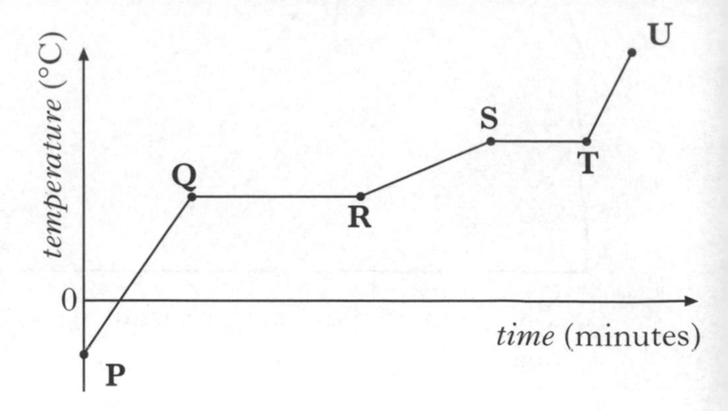

During the time interval QR, which of the following statements is/are correct?

I There is a change in the state of the substance.

II The substance changes state from a liquid to a gas.

III Heat is absorbed by the substance.

A I only

B III only

C I and II only

D I and III only

E I, II and III

8. A student writes the following statements about electrical conductors.

I Only protons are free to move.

II Only electrons are free to move.

III Only negative charges are free to move.

Which of the statements is/are correct?

A I only

B II only

C III only

D I and II only

E II and III only

9. A charge of 15 C passes through a resistor in 12 s. The potential difference across the resistor is 6 V.

The power developed by the resistor is

A 4·8 W

B 7·5 W

C 9·4 W

D 30 W

E 1080 W.

10. A circuit is set up as shown.

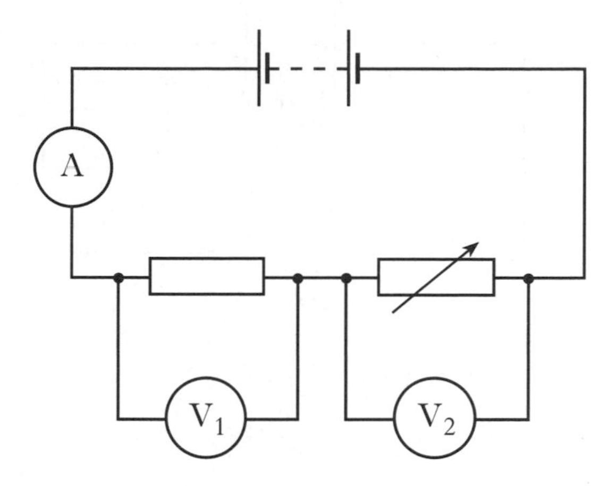

The resistance of the variable resistor is increased.

Which row in the table shows the effect on the readings on the ammeter and voltmeters?

	Reading on ammeter	*Reading on voltmeter* V_1	*Reading on voltmeter* V_2
A	decreases	decreases	decreases
B	increases	unchanged	increases
C	decreases	increases	decreases
D	increases	unchanged	decreases
E	decreases	decreases	increases

[Turn over

11. A circuit is set up as shown.

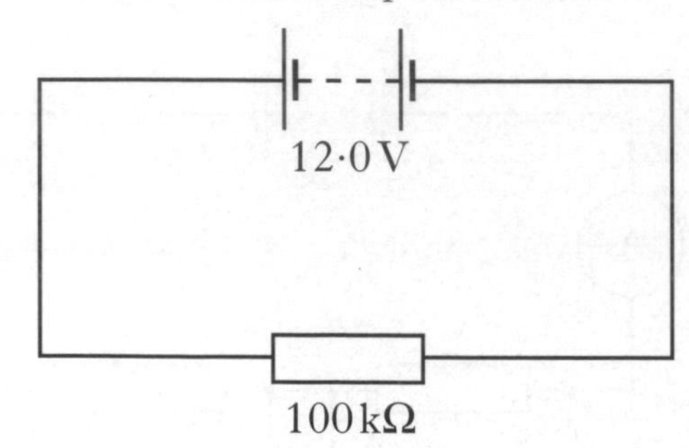

The power supplied to the resistor is

A $1{\cdot}20 \times 10^{-4}$ W

B $1{\cdot}44 \times 10^{-3}$ W

C 1·44 W

D 694 W

E $1{\cdot}20 \times 10^{6}$ W.

12. Which of the following devices transforms light energy into electrical energy?

A LED

B Thermocouple

C Microphone

D Solar cell

E Transistor

13. Which of the following is the correct symbol for an n-channel enhancement MOSFET?

A

B

C

D

E

t

14. Which of the following is an example of a longitudinal wave?

A Light wave

B Infra-red wave

C Radio wave

D Sound wave

E Water wave

15. The diagram shows a list of the members of the electromagnetic spectrum in order of increasing wavelength.

gamma rays	**P**	ultraviolet	**Q**	infrared	**R**	TV and Radio

Which row in the table shows the radiation represented by the letters **P**, **Q** and **R**?

	P	**Q**	**R**
A	microwaves	visible light	x-rays
B	visible light	microwaves	x-rays
C	x-rays	visible light	microwaves
D	visible light	x-rays	microwaves
E	x-rays	microwaves	visible light

16. The diagram shows what happens to a ray of light when it strikes a glass block.

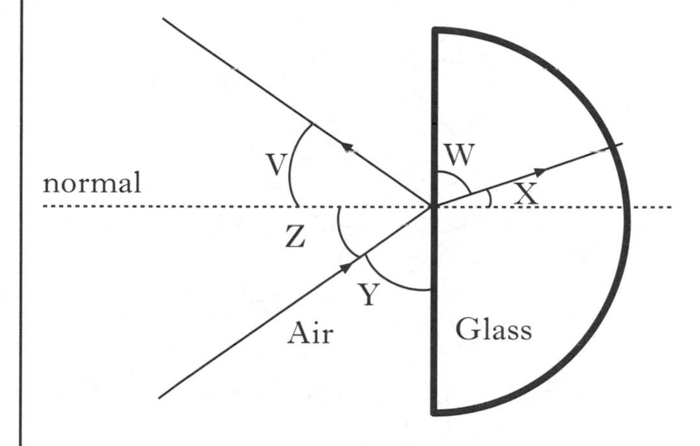

Which row in the table identifies the angle of incidence and the angle of refraction?

	Angle of Incidence	*Angle of Refraction*
A	V	W
B	Y	W
C	Y	X
D	Z	W
E	Z	X

[Turn over

17. The diagram below shows a simple model of an atom.

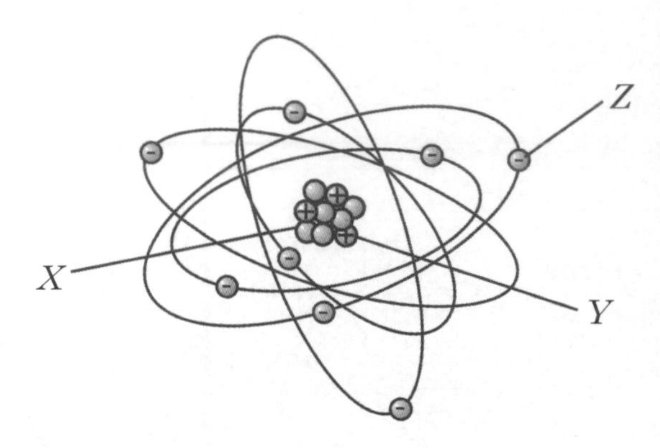

Which row in the table identifies particles *X*, *Y* and *Z*?

	X	*Y*	*Z*
A	electron	proton	neutron
B	proton	neutron	electron
C	neutron	electron	proton
D	electron	neutron	proton
E	neutron	proton	electron

18. A student makes the following statements about ionising radiations.

I Ionisation occurs when an atom loses an electron.

II Gamma radiation produces greater ionisation (density) than alpha particles.

III An alpha particle consists of 2 protons, 2 neutrons and 2 electrons.

Which of the statements is/are correct?

A I only

B II only

C I and II only

D II and III only

E I, II and III

19. A sample of tissue has a mass of 0·05 kg.

The tissue is exposed to radiation and absorbs 0·1 J of energy in 2 minutes.

The absorbed dose is

A 0·005 Gy

B 0·1 Gy

C 0·5 Gy

D 2 Gy

E 6 Gy.

20. During fission, a neutron splits a uranium nucleus into two nuclei, X and Y, as shown below.

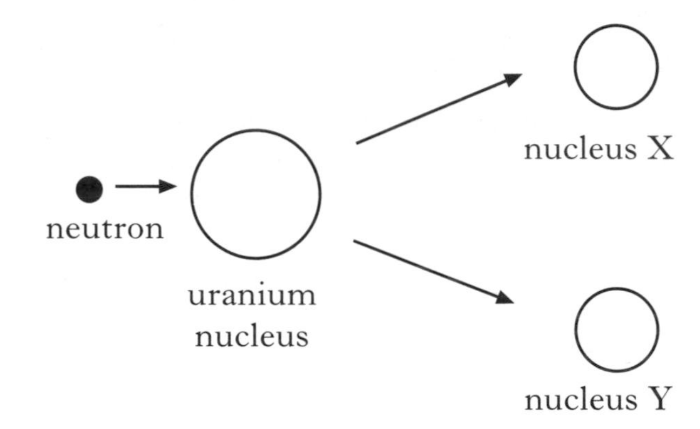

For a chain reaction to occur which of the following **must** also be released?

A Protons

B Electrons

C Neutrons

D Alpha particles

E Gamma radiation

Candidates are reminded that the answer sheet for Section A MUST be placed INSIDE the front cover of the answer book.

SECTION B

Marks

Write your answers to questions 21–29 in the answer book.

All answers must be written clearly and legibly in ink.

21. A ski lift with a gondola of mass 2000 kg travels to a height of 540 m from the base station to a station at the top of the mountain.

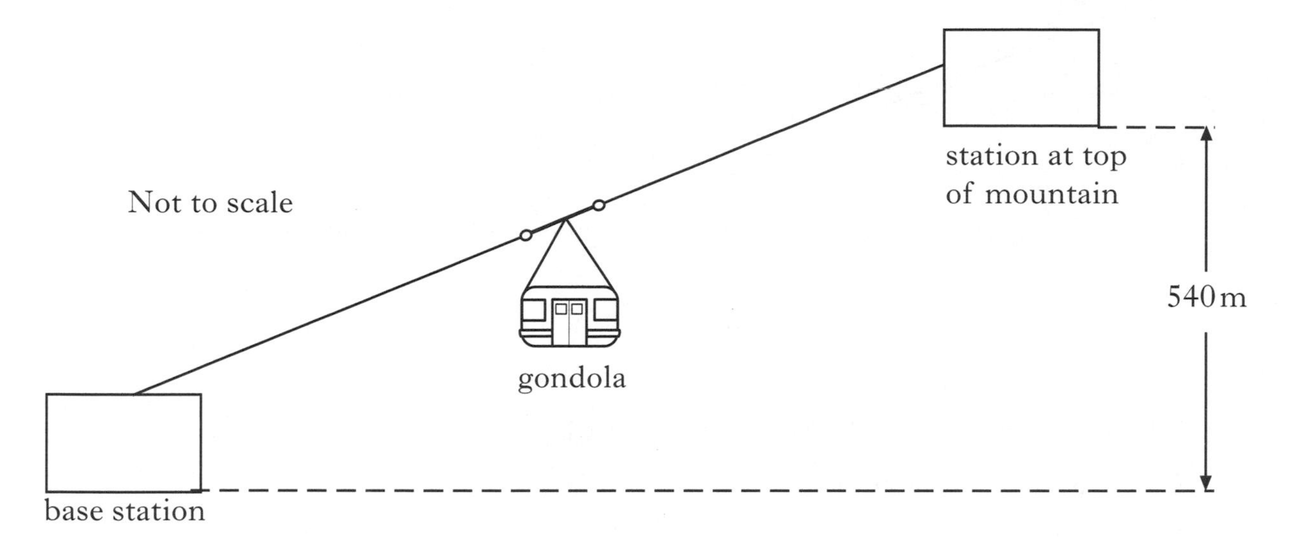

(*a*) Calculate the gain in gravitational potential energy of the gondola. **2**

(*b*) During the journey, the kinetic energy of the gondola is 64 000 J.

Calculate the speed of the gondola. **2**

(*c*) The ski lift requires a motor which operates at 380 V to take the gondola up the mountain. The maximum power produced is 45·6 kW.

(i) Calculate the maximum current in the motor. **2**

(ii) Calculate the electrical energy used by the motor when it has been operating at its maximum power for a total time of 1 hour. **2**

(8)

[Turn over

Marks

22. A child sledges down a hill.

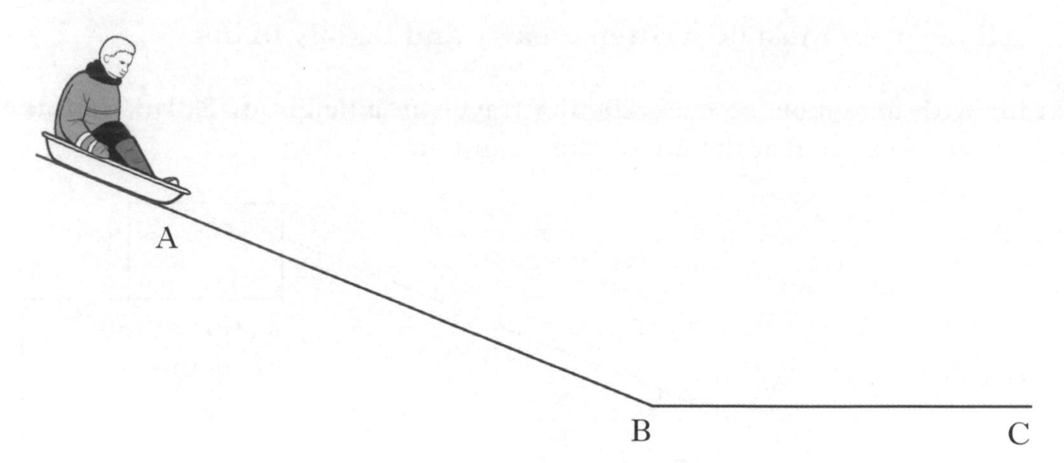

The sledge and child are released from rest at point A. They reach a speed of 3 m/s at point B.

(*a*) The sledge and child take 5 s to reach point B.

Calculate the acceleration. **2**

(*b*) The sledge and child have a combined mass of 40 kg.

Calculate the unbalanced force acting on them. **2**

(*c*) After the sledge and child pass point B, they slow down, coming to a halt at point C.

Explain this motion in terms of forces. **2**

(6)

Marks

23. The following apparatus is used to determine the speed of a pellet as it leaves an air rifle. The air rifle fires a pellet into the plasticine, causing the vehicle to move.

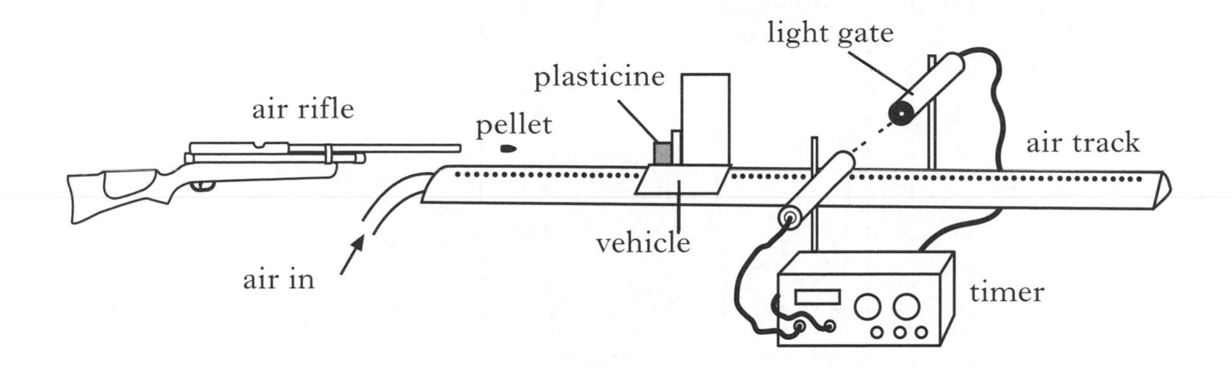

(*a*) Describe how the apparatus is used to determine the speed of the vehicle.

Your description must include:

- the measurements made
- any necessary calculations. **2**

(*b*) The speed of the vehicle is calculated as 0·35 m/s after impact.

The mass of the pellet is $5{\cdot}0 \times 10^{-4}$ kg. The mass of the vehicle and plasticine before impact is 0·30 kg.

(i) Show that the momentum of the pellet **before** impact with the plasticine is 0·105 kg m/s. **1**

(ii) Hence, calculate the velocity of the pellet **before** impact with the plasticine. **1**

(*c*) At a firing range a pellet is fired horizontally at a target 40 m away. It takes 0·20 s to reach the target.

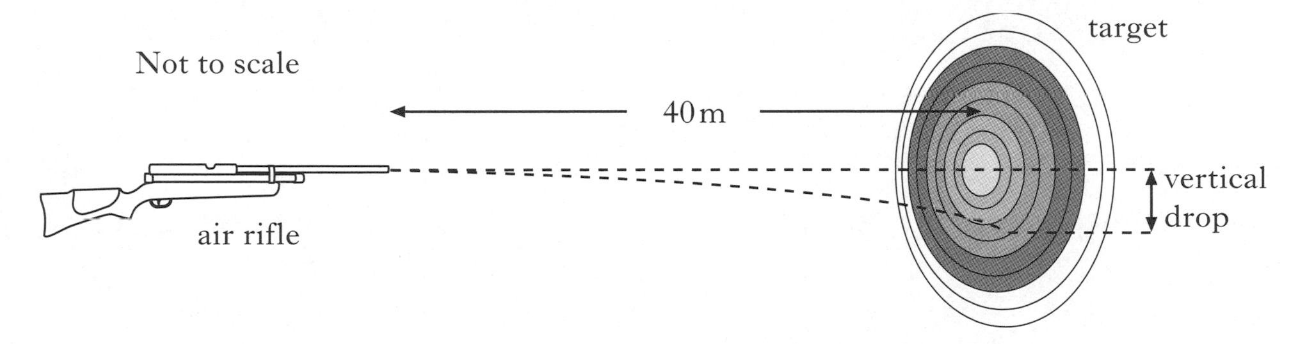

(i) Calculate the **vertical** velocity of the pellet on reaching the target. **2**

(ii) Calculate the vertical drop. **2**

(8)

[Turn over

Marks

24. A fridge/freezer has water and ice dispensers as shown.

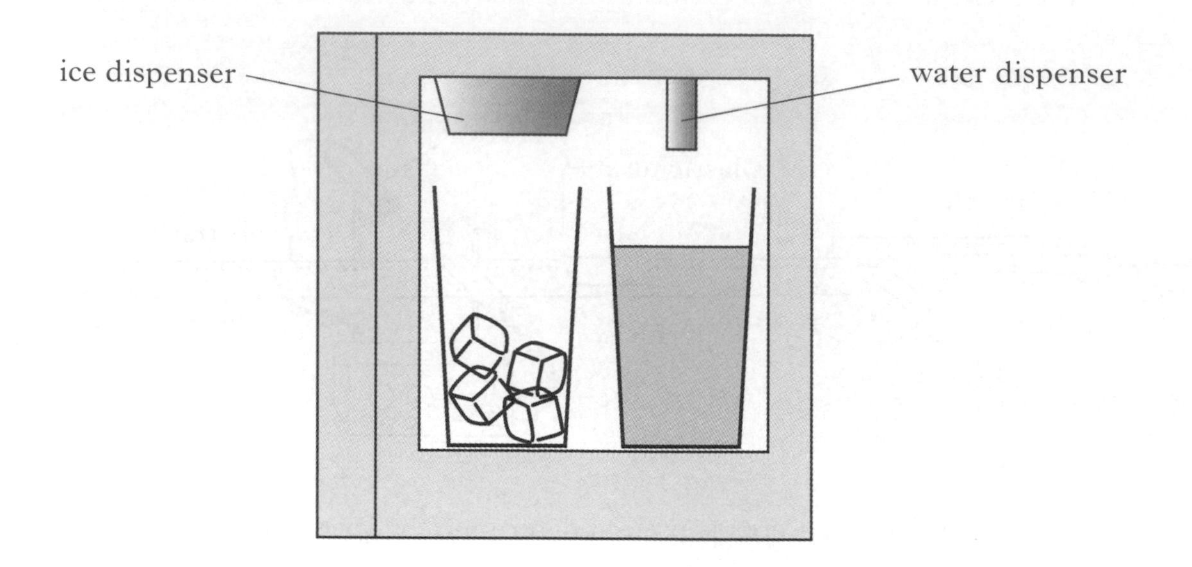

(*a*) Water of mass 0·1 kg flows into the freezer at 15 °C and is cooled to 0 °C. Calculate the energy removed when the water cools. **2**

(*b*) Calculate how much energy is released when 0·1 kg of water at 0 °C changes to 0·1 kg of ice at 0 °C. **2**

(*c*) The fridge/freezer system removes heat energy at a rate of 125 J/s.

(i) Calculate the minimum time taken to produce 0·1 kg of ice from 0·1 kg of water at 15 °C. **3**

(ii) Explain why the actual time taken to make the ice will be longer than the value calculated in part (i). **2**

(9)

Marks

25. A student sets up the following circuit to investigate the resistance of resistor R.

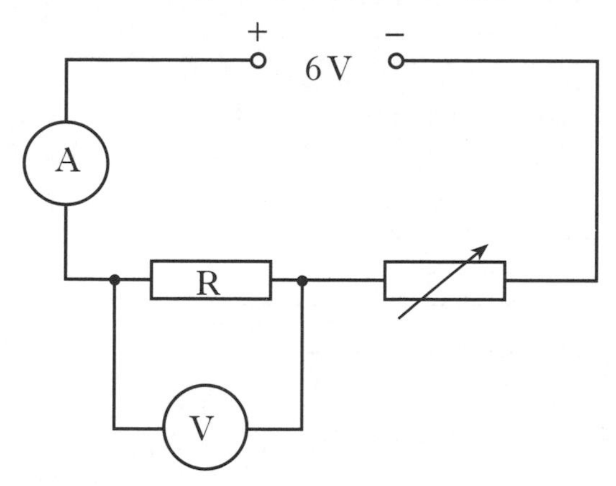

The variable resistor is adjusted and the voltmeter and ammeter readings are noted. The following graph is obtained from the experimental results.

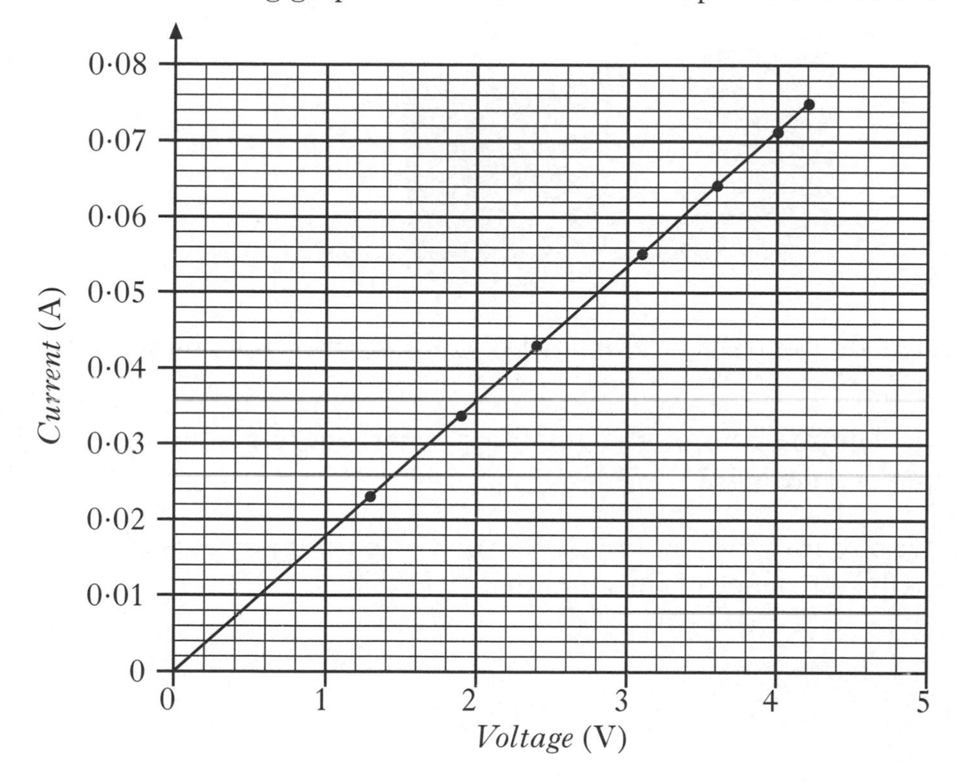

(*a*) (i) Calculate the value of the resistor R when the reading on the voltmeter is 4·2 V. **3**

(ii) Using information from the graph, state whether the resistance of the resistor R, **increases**, **stays the same** or **decreases** as the voltage increases.

Justify your answer. **2**

(*b*) The student is given a task to combine two resistors from a pack containing one each of 33 Ω, 56 Ω, 82 Ω, 150 Ω, 270 Ω, 390 Ω.

Show by calculation which **two** resistors should be used to give:

(i) the largest combined resistance; **2**

(ii) the smallest combined resistance. **2**

(9)

Marks

26. An MP3 player is charged from the mains supply of 230 V using a transformer, which has an output voltage of 5 V and an output current of 1 A.

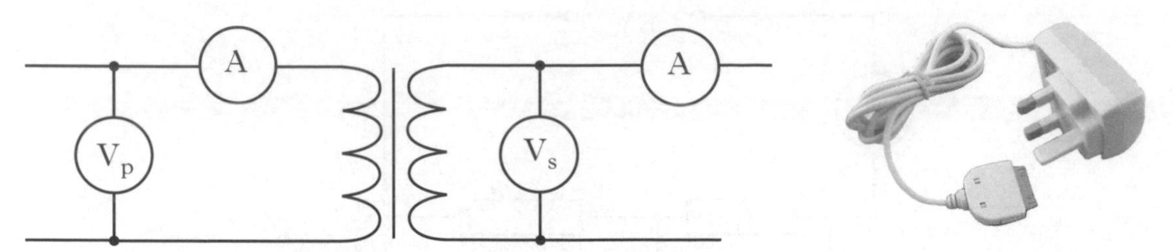

circuit diagram of transformer

MP3 Charger

(*a*) Calculate the current in the primary circuit. **2**

(*b*) The MP3 player is then put on a docking station with external speakers.

(i) Calculate the resistance of a 10 W speaker when the voltage across it is 9 V. **2**

(ii) Calculate the gain of the amplifier in the docking station when the input voltage is 1·5 V. **2**

(*c*) The input power to the amplifier is 25 W. The output power is 20 W. Calculate the efficiency of the amplifier. **2**

(8)

Marks

27. A student is short sighted.

(*a*) (i) What does the term "short sighted" mean? **1**

(ii) What type of lens is required to correct this eye defect? **1**

(iii) The focal length of the lens needed to correct the student's short sight is 180 mm. Calculate the power of this lens. **2**

(*b*) In the eye, refraction of light occurs at both the cornea and the lens. Some eye defects can be corrected using a laser. Light from the laser is used to change the shape of the cornea.

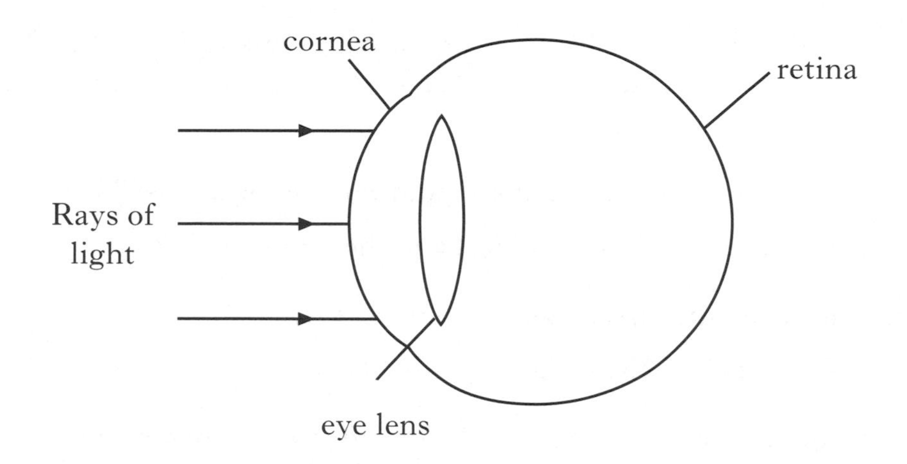

(i) State what is meant by refraction of light. **1**

(ii) The laser emits light of wavelength 7×10^{-7} m.

Calculate the frequency of the light. **2**

(*c*) Lasers can be used in optical fibres for medical purposes.

(i) Copy and complete the path of the laser light along the optical fibre. **2**

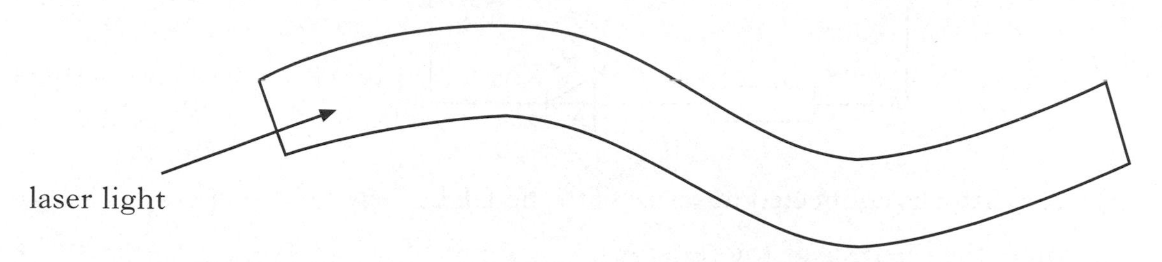

(ii) Name the effect when the laser light hits the inside surface of the fibre. **1**

(10)

[Turn over

Marks

28. Parking sensors are fitted to the rear bumper of some cars. A buzzer emits audible beeps, which become more frequent as the car moves closer to an object.

Ultrasonic pulses are emitted from the rear of the car. Objects behind the car reflect the pulses, which are detected by sensors. Ultrasonic pulses travel at the speed of sound.

(*a*) The time between these pulses being sent and received is 2×10^{-3} s.

Calculate the distance between the object and the rear of the car. **3**

(*b*) At a certain distance, the buzzer beeps every 0·125 s.

Calculate the frequency of the beeps. **2**

(*c*) The sensor operates at a voltage of 12 V and has a current range of 20–200 mA.

Calculate the maximum power rating of the sensor. **3**

(*d*) An LED system can be added so that it flashes at the same frequency as the beeps from the buzzer. The LED circuit is shown below.

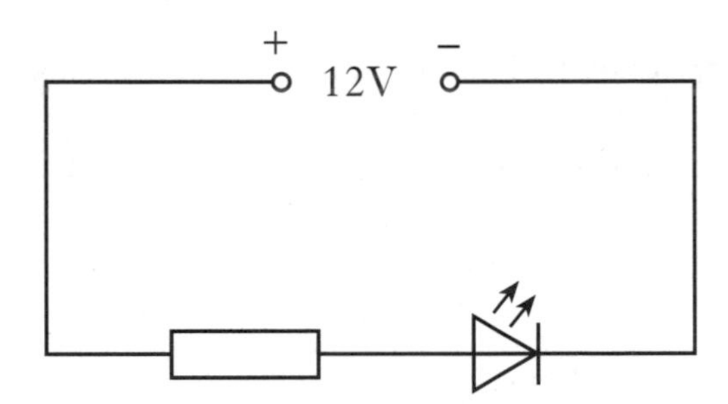

(i) A resistor is connected in series with the LED.

State the purpose of the resistor. **1**

(ii) When lit, the LED has a voltage of 3·5 V across it and a current of 200 mA.

Calculate the value of the resistor. **3**

(12)

Marks

29. A radioactivity kit includes three radioactive sources each made up as shown.

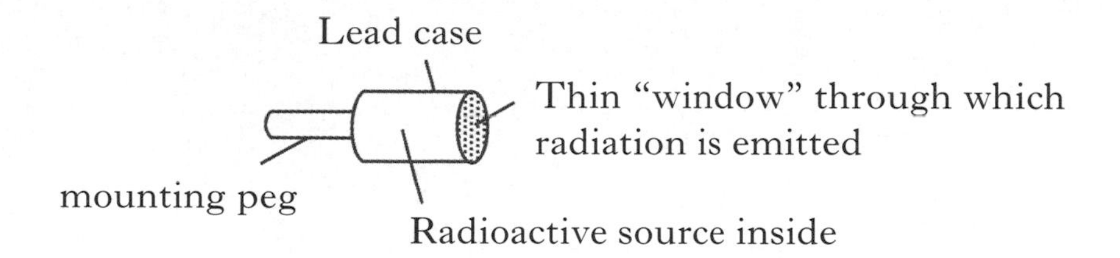

Information about these sources is given in the table below.

	Radiation Emitted	*Radioactive Element*
Source 1	Alpha	Americium 241
Source 2	Beta	Strontium 90
Source 3	Gamma + Beta	Cobalt 60

(*a*) (i) Describe an experiment to show which is the alpha emitting source.

Your description must include:

- equipment used
- measurements taken
- an explanation of the results. **3**

(ii) The radioactive material in Source 3 emits both beta and gamma radiations. Describe how the window of the casing could be modified so that the beta radiation is stopped. **1**

(*b*) Strontium 90 has a half life of 28 years. Calculate how many years it takes for the activity to decrease to 1/16th of its original value. **2**

(*c*) (i) A technician working with Source 1 receives an absorbed dose of 20 μGy of alpha particles. Calculate the total equivalent dose received by the technician. **2**

(ii) Describe two ways in which the technician could reduce his absorbed dose. **2**

(10)

[*END OF QUESTION PAPER*]

[BLANK PAGE]

INTERMEDIATE 2

2010

[BLANK PAGE]

X069/201

NATIONAL QUALIFICATIONS 2010

FRIDAY, 28 MAY
1.00 PM – 3.00 PM

PHYSICS
INTERMEDIATE 2

Read Carefully

Reference may be made to the Physics Data Booklet

1 All questions should be attempted.

Section A (questions 1 to 20)

2 Check that the answer sheet is for Physics Intermediate 2 (Section A).

3 For this section of the examination you must use an **HB pencil** and, where necessary, an eraser.

4 Check that the answer sheet you have been given has **your name**, **date of birth**, **SCN** (Scottish Candidate Number) and **Centre Name** printed on it.

Do not change any of these details.

5 If any of this information is wrong, tell the Invigilator immediately.

6 If this information is correct, **print** your name and seat number in the boxes provided.

7 There is **only one correct** answer to each question.

8 Any rough working should be done on the question paper or the rough working sheet, **not** on your answer sheet.

9 At the end of the exam, put the **answer sheet for Section A inside the front cover of your answer book**.

10 Instructions as to how to record your answers to questions 1–20 are given on page three.

Section B (questions 21 to 30)

11 Answer the questions numbered 21 to 30 in the answer book provided.

12 **All answers must be written clearly and legibly in ink.**

13 Fill in the details on the front of the answer book.

14 Enter the question number clearly in the margin of the answer book beside each of your answers to questions 21 to 30.

15 Care should be taken to give an appropriate number of significant figures in the final answers to calculations.

DATA SHEET

Speed of light in materials

Material	*Speed* in m/s
Air	$3{\cdot}0 \times 10^8$
Carbon dioxide	$3{\cdot}0 \times 10^8$
Diamond	$1{\cdot}2 \times 10^8$
Glass	$2{\cdot}0 \times 10^8$
Glycerol	$2{\cdot}1 \times 10^8$
Water	$2{\cdot}3 \times 10^8$

Speed of sound in materials

Material	*Speed* in m/s
Aluminium	5200
Air	340
Bone	4100
Carbon dioxide	270
Glycerol	1900
Muscle	1600
Steel	5200
Tissue	1500
Water	1500

Gravitational field strengths

	Gravitational field strength on the surface in N/kg
Earth	10
Jupiter	26
Mars	4
Mercury	4
Moon	1·6
Neptune	12
Saturn	11
Sun	270
Venus	9

Specific heat capacity of materials

Material	*Specific heat capacity* in J/kg °C
Alcohol	2350
Aluminium	902
Copper	386
Glass	500
Ice	2100
Iron	480
Lead	128
Oil	2130
Water	4180

Specific latent heat of fusion of materials

Material	*Specific latent heat of fusion* in J/kg
Alcohol	$0{\cdot}99 \times 10^5$
Aluminium	$3{\cdot}95 \times 10^5$
Carbon dioxide	$1{\cdot}80 \times 10^5$
Copper	$2{\cdot}05 \times 10^5$
Iron	$2{\cdot}67 \times 10^5$
Lead	$0{\cdot}25 \times 10^5$
Water	$3{\cdot}34 \times 10^5$

Melting and boiling points of materials

Material	*Melting point* in °C	*Boiling point* in °C
Alcohol	−98	65
Aluminium	660	2470
Copper	1077	2567
Glycerol	18	290
Lead	328	1737
Iron	1537	2747

Specific latent heat of vaporisation of materials

Material	*Specific latent heat of vaporisation* in J/kg
Alcohol	$11{\cdot}2 \times 10^5$
Carbon dioxide	$3{\cdot}77 \times 10^5$
Glycerol	$8{\cdot}30 \times 10^5$
Turpentine	$2{\cdot}90 \times 10^5$
Water	$22{\cdot}6 \times 10^5$

Radiation weighting factors

Type of radiation	*Radiation weighting factor*
alpha	20
beta	1
fast neutrons	10
gamma	1
slow neutrons	3

SECTION A

For questions 1 to 20 in this section of the paper the answer to each question is either A, B, C, D or E. Decide what your answer is, then, using your pencil, put a horizontal line in the space provided—see the example below.

EXAMPLE

The energy unit measured by the electricity meter in your home is the

A kilowatt-hour

B ampere

C watt

D coulomb

E volt.

The correct answer is **A**—kilowatt-hour. The answer **A** has been clearly marked in **pencil** with a horizontal line (see below).

A B C D E

Changing an answer

If you decide to change your answer, carefully erase your first answer and, using your pencil, fill in the answer you want. The answer below has been changed to **E**.

A B C D E

[Turn over

SECTION A

Answer questions 1–20 on the answer sheet.

1. Which of the following is a scalar quantity?

A Force

B Acceleration

C Momentum

D Velocity

E Energy

2. A student investigates the speed of a trolley as it moves down a slope.

The apparatus is set up as shown.

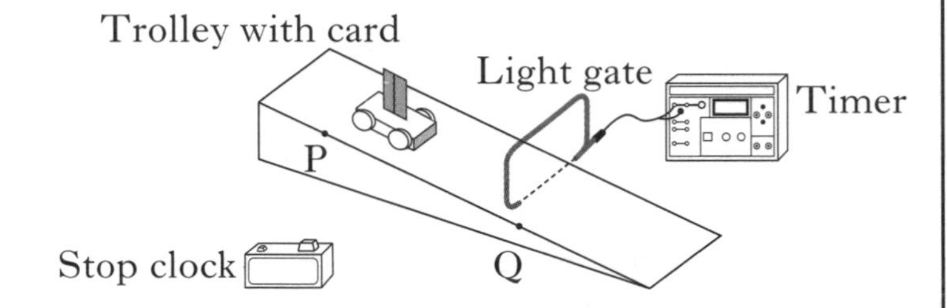

The following measurements are recorded.

distance from P to Q = 1·0 m
length of card on trolley = 0·04 m
time taken for trolley to travel from P to Q = 2·5 s
time taken for card to pass through light gate = 0·05 s

The speed at Q is

A 0·002 m/s

B 0·016 m/s

C 0·40 m/s

D 0·80 m/s

E 20 m/s.

3. Two forces, each of 7 N, act on an object O.

The forces act as shown.

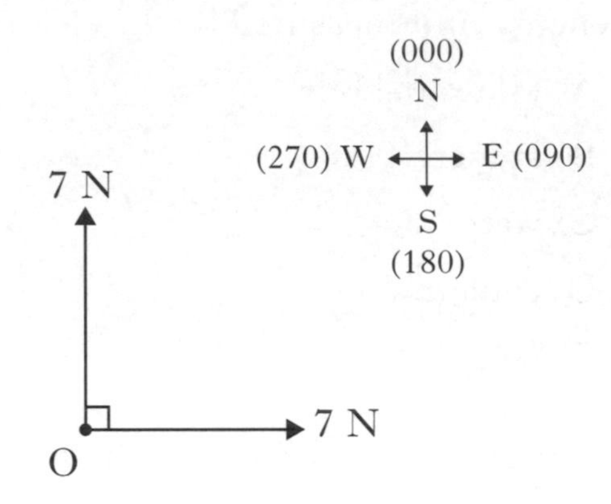

The resultant of these two forces is

A 7 N at a bearing of 135

B 9·9 N at a bearing of 045

C 9·9 N at a bearing of 135

D 14 N at a bearing of 045

E 14 N at a bearing of 135.

4 A package is released from a helicopter flying horizontally at a constant speed of 40 m/s.

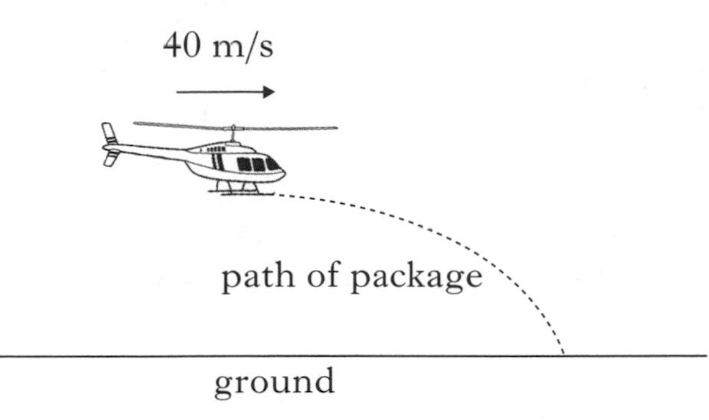

The package takes 3·0 s to reach the ground.

The effects of air resistance can be ignored.

Which row in the table shows the horizontal speed and vertical speed of the package just before it hits the ground?

	Horizontal speed (m/s)	*Vertical speed* (m/s)
A	0	30
B	30	30
C	30	40
D	40	30
E	40	40

5. 100 g of a solid is heated by a 50 W heater. The graph of temperature of the substance against time is shown.

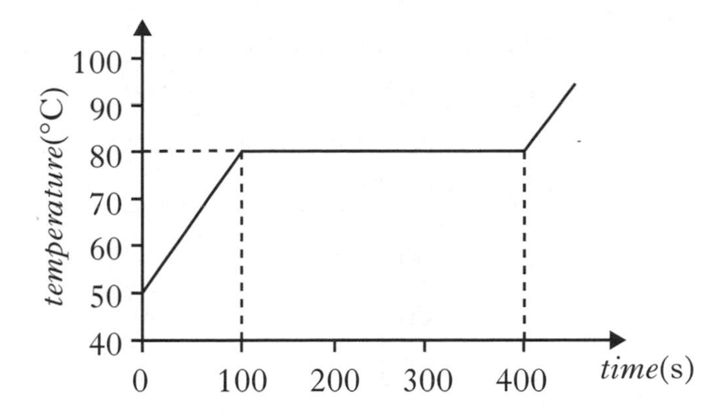

The specific latent heat of fusion of the substance is

A $1{\cdot}3 \times 10^3$ J/kg

B $1{\cdot}5 \times 10^3$ J/kg

C $3{\cdot}0 \times 10^3$ J/kg

D $1{\cdot}5 \times 10^5$ J/kg

E $1{\cdot}9 \times 10^5$ J/kg.

[Turn over

6. A crate of mass 200 kg is pushed a distance of 20 m across a level floor.

The crate is pushed with a force of 150 N.

The force of friction acting on the crate is 50 N.

The work done in pushing the crate across the floor is

A 1000 J

B 2000 J

C 3000 J

D 4000 J

E 20 000 J.

7. A student makes the following statements about electrical circuits.

I The sum of the potential differences across components connected in series is equal to the supply voltage.

II The sum of the currents in parallel branches is equal to the current drawn from the supply.

III The potential difference across components connected in parallel is the same for each component.

Which of the statements is/are correct?

A I only

B III only

C I and II only

D II and III only

E I, II and III

8. Three resistors are connected as shown

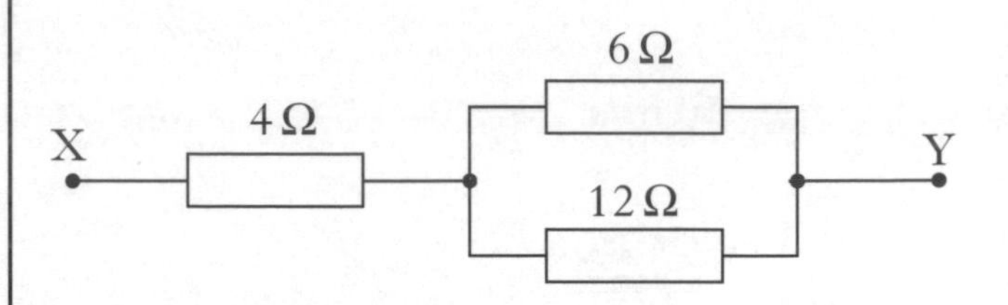

The total resistance between X and Y is

A 2 Ω

B 4 Ω

C 8 Ω

D 13 Ω

E 22 Ω.

9. The resistance of a wire is 6 Ω.

The current in the wire is 2 A.

The power developed in the wire is

A 3 W

B 12 W

C 18 W

D 24 W

E 72 W.

10. The voltage of the mains supply in the UK is 230 V a.c.

Which row in the table shows the peak voltage and frequency of the mains supply in the UK?

	peak voltage (V)	*frequency* (Hz)
A	175	50
B	175	60
C	230	50
D	325	50
E	325	60

11. The diagram shows a model bicycle dynamo.

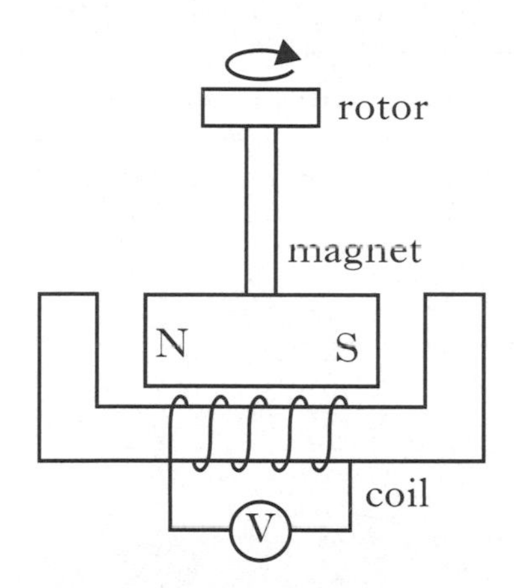

When the rotor is turned the magnet rotates, inducing a voltage in the coil. The induced voltage can be decreased by

A increasing the number of turns on the coil

B decreasing the number of turns on the coil

C using a stronger magnet

D turning the rotor faster

E reversing the direction of rotation of the magnet.

12. The graph below shows how the input voltage V_I to an amplifier varies with time t.

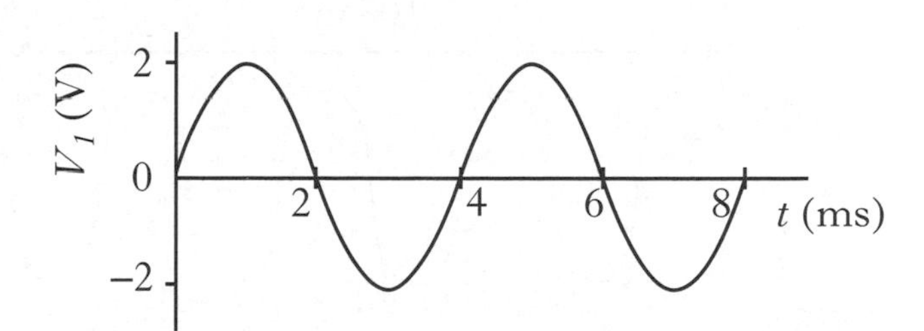

The amplifier has a voltage gain of 10.

Which graph shows how the output voltage V_0 of the amplifier varies with time t?

A

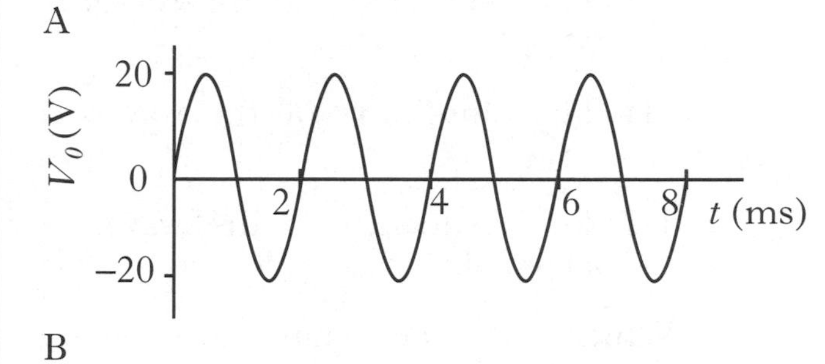

B

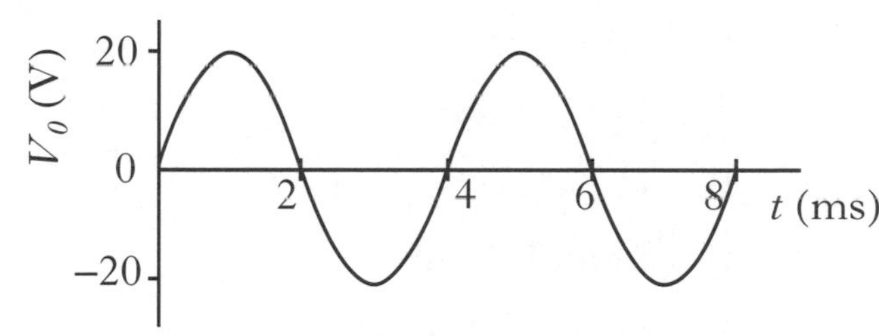

C

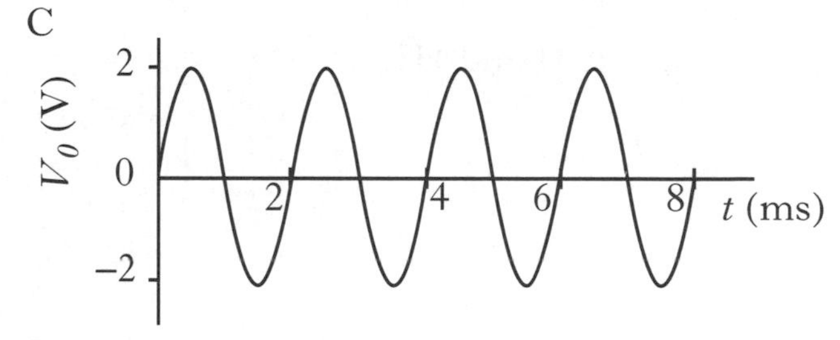

D

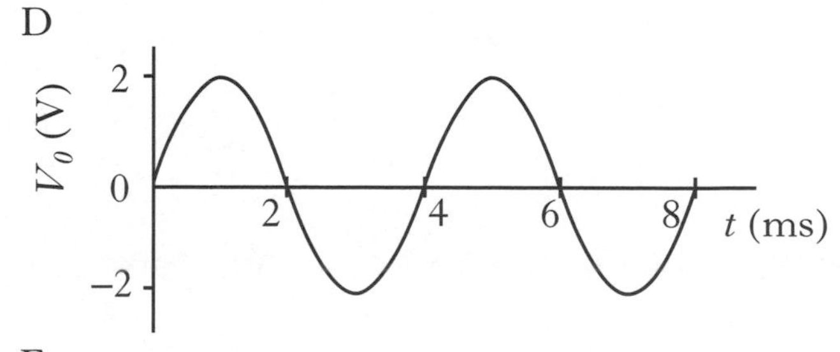

E

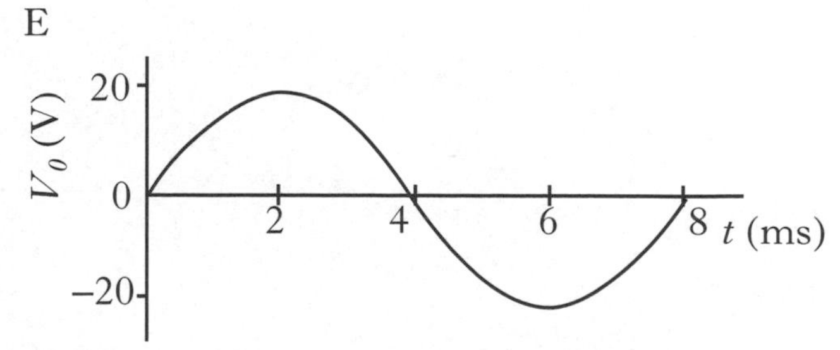

13. The diagram gives information about a wave.

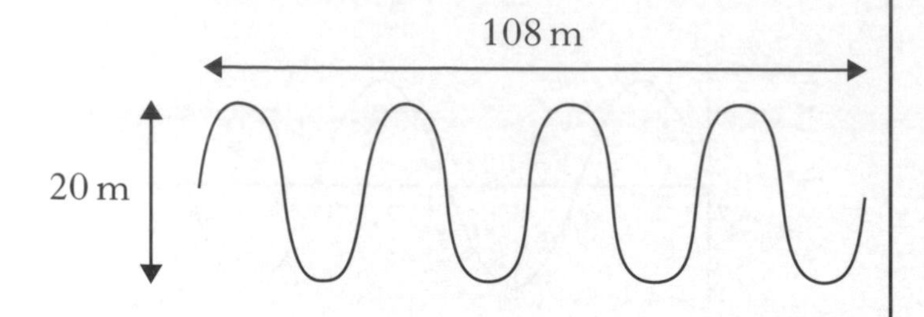

The time taken for the waves to travel 108 m is 0·5 s.

A student makes the following statements about the waves.

I The wavelength of the waves is 27 m.

II The amplitude of the waves is 20 m.

III The frequency of the waves is 8 Hz.

Which of the statements is/are correct?

A I only

B II only

C I and III only

D II and III only

E I, II and III

14. The diagram shows members of the electromagnetic spectrum in order of increasing wavelength.

Gamma rays	P	Ultraviolet radiation	Q	Infrared radiation	R	TV & radio waves

increasing wavelength →

Which row in the table identifies the radiations represented by the letters P, Q and R?

	P	*Q*	*R*
A	X-rays	visible light	microwaves
B	X-rays	microwaves	visible light
C	microwaves	visible light	X-rays
D	visible light	microwaves	X-rays
E	visible light	X-rays	microwaves

15. An object is placed in front of a converging lens as shown.

The position of the image formed by the lens is also shown.

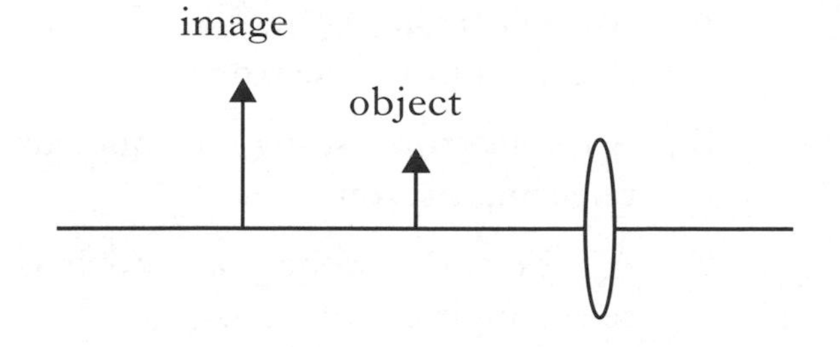

The focal length of the lens is 100 mm.

The distance between the lens and the object is

A 50 mm

B 100 mm

C 150 mm

D 200 mm

E 250 mm.

16. A converging lens has a focal length of 50 mm.

The power of the lens is

A +0·02 D

B +0·2 D

C −0·2 D

D +20 D

E −20 D.

17. A student makes the following statements about a carbon atom.

I The atom is made up only of protons and neutrons.

II The nucleus of the atom contains protons, neutrons and electrons.

III The nucleus of the atom contains only protons and neutrons.

Which of the statements is/are correct?

A I only

B II only

C III only

D I and II only

E I and III only

18. Human tissue can be damaged by exposure to radiation.

On which of the following factors does the risk of biological harm depend?

I The absorbed dose.

II The type of radiation.

III The body organs or tissue exposed.

A I only

B I and II only

C II only

D II and III only

E I, II and III

[Turn over

19. Information about a radioactive source is given in Table 1.

Table 1

Activity	*Energy absorbed per kilogram of tissue*	*Radiation weighting factor*
500 MBq	0·2 μJ	10

Which row in Table 2 gives the correct information for the radioactive source?

Table 2

	Absorbed Dose	*Equivalent Dose*
A	0·2 μGy	2 μSv
B	500 MGy	10 Sv
C	10 Gy	0·2 μSv
D	20 μGy	50 MSv
E	2 μGy	0·2 μSv

20. In a nuclear reactor a chain reaction releases energy from nuclei.

Which of the following statements describes the beginning of a chain reaction?

A An electron splits a nucleus releasing more electrons.

B An electron splits a nucleus releasing protons.

C A proton splits a nucleus releasing more protons.

D A neutron splits a nucleus releasing electrons.

E A neutron splits a nucleus releasing more neutrons.

Candidates are reminded that the answer sheet for Section A MUST be placed INSIDE the front cover of the answer book.

SECTION B

Marks

Write your answers to questions 21–30 in the answer book.

All answers must be written clearly and legibly in ink.

21. A balloon of mass 400 kg rises vertically from the ground.

The graph shows how the vertical speed of the balloon changes during the first 100 s of its upward flight.

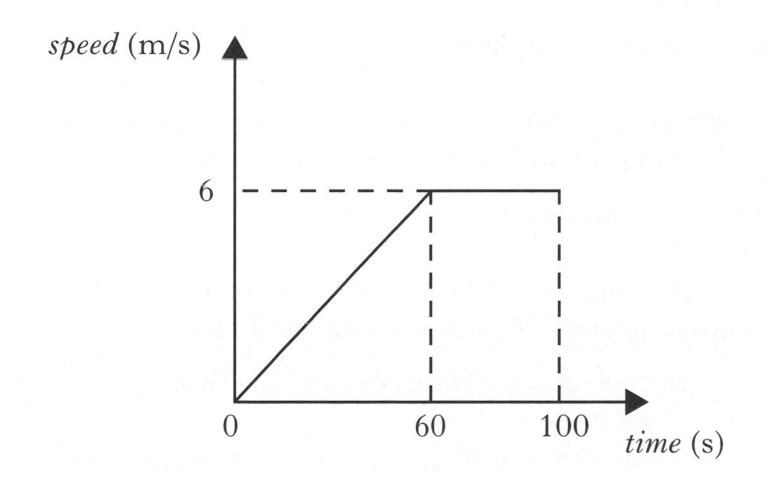

(*a*) Calculate the acceleration of the balloon during the first 60 s. **2**

(*b*) Calculate the distance travelled by the balloon in 100 s. **2**

(*c*) Calculate the average speed of the balloon during the first 100 s. **2**

(*d*) Calculate the weight of the balloon. **2**

(*e*) Calculate the total upward force acting on the balloon during the first 60 s of its flight. **3**

(11)

Marks

22. Inside a storm cloud water droplets move around and collide with each other.

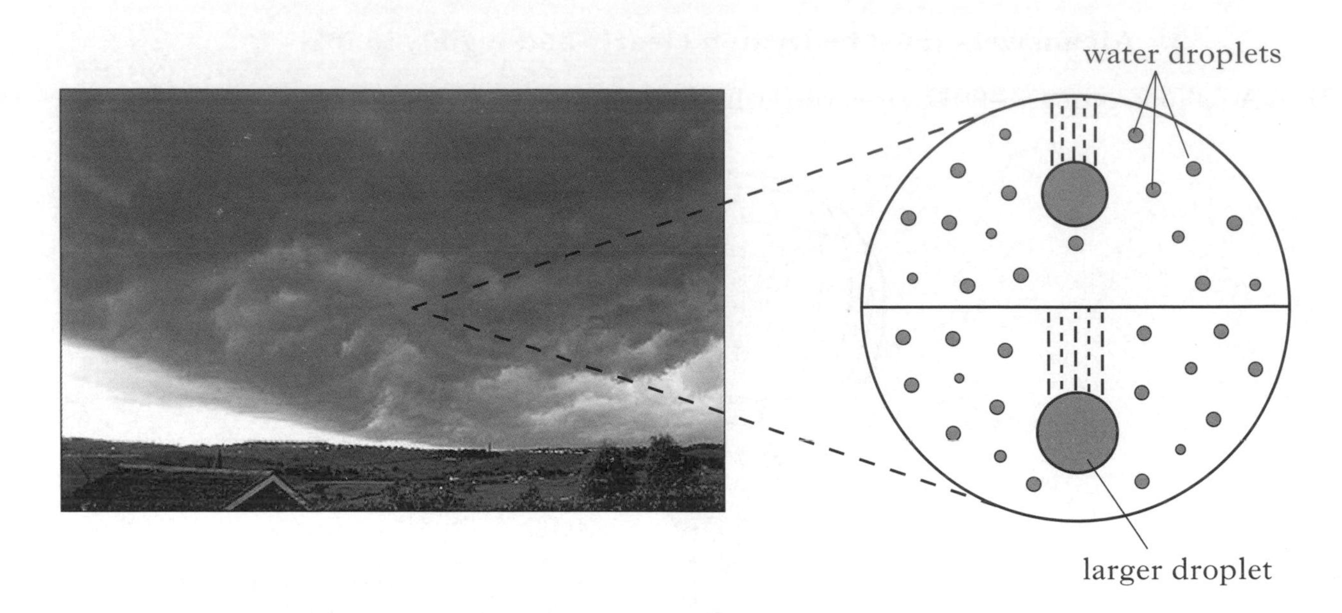

(*a*) A water droplet of mass 2·0 g moving at a speed of 4·0 m/s collides with a stationary water droplet of mass 1·2 g. The two droplets join together to form a larger droplet.

Calculate the speed of this larger droplet after the collision. **2**

(*b*) Another water droplet within the cloud is falling with a constant speed. Draw a diagram showing the forces acting on this droplet.

Name these forces and show their directions. **2**

(*c*) The motion of water droplets in the cloud causes flashes of lightning. One lightning flash transfers 1650 C of charge in 0·15 s.

Calculate the electric current produced by this flash. **2**

(*d*) Why does an observer, standing 3 km from a thunder cloud, see a lightning flash before he hears the thunder? **1**

(7)

Marks

23. On the planet Mercury the surface temperature at night is −173 °C. The surface temperature during the day is 307 °C. A rock lying on the surface of the planet has a mass of 60 kg.

(*a*) The rock absorbs $2{\cdot}59 \times 10^7$ J of heat energy from the Sun during the day.

Calculate the specific heat capacity of the rock. **2**

(*b*) Heat is released at a steady rate of 1440 J/s at night.

Calculate the time taken for the rock to release $2{\cdot}59 \times 10^7$ J of heat. **2**

(*c*) Energy from these rocks could be used to heat a base on the surface of Mercury.

How many 60 kg rocks would be needed to supply a 288 kW heating system? **2**

(*d*) Using information from the data sheet, would it be **easier**, **the same** or **more difficult** to lift rocks on Mercury compared to Earth?

You **must** explain your answer. **2**

(8)

[Turn over

Marks

24. A student sets up the following circuit.

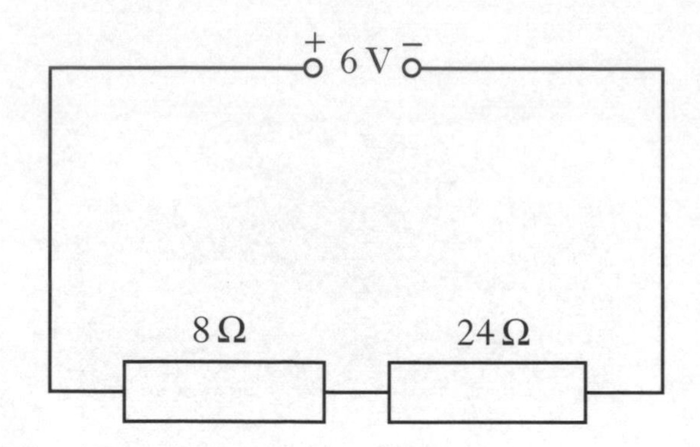

(*a*) Calculate the current in the 8 Ω resistor. **3**

(*b*) Calculate the voltage across the 8 Ω resistor. **2**

(*c*) The 24 Ω resistor is replaced by one of **greater** resistance. How will this affect the voltage across the 8 Ω resistor?

Explain your answer. **2**

(7)

Marks

25. In a lab experiment a technician builds a transformer and uses electrical meters to take a number of measurements, as shown in the diagram.

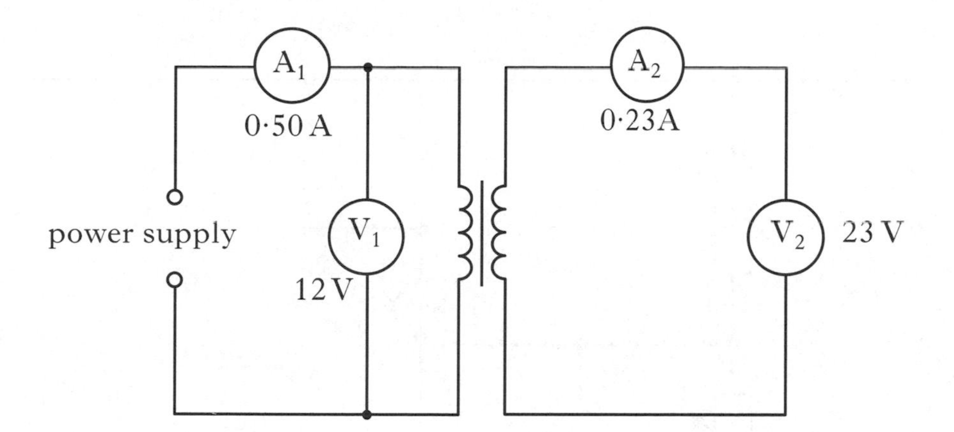

(*a*) The technician has a choice of an a.c. or a d.c. power supply. Which power supply should be used?

Explain your answer. **2**

(*b*) Calculate the electrical power in the primary circuit of the transformer. **2**

(*c*) Calculate the electrical power in the secondary circuit of the transformer. **1**

(*d*) Calculate the percentage efficiency of the transformer. **2**

(*e*) Another experiment uses a different transformer. It is 100% efficient. The primary coil has 1500 turns and the secondary coil contains 3000 turns.

Calculate the secondary voltage when the primary voltage is 12 V. **2**

(9)

[Turn over

Marks

26. Water in a fish tank has to be maintained at a constant temperature. Part of the electronic circuit which controls the temperature is shown.

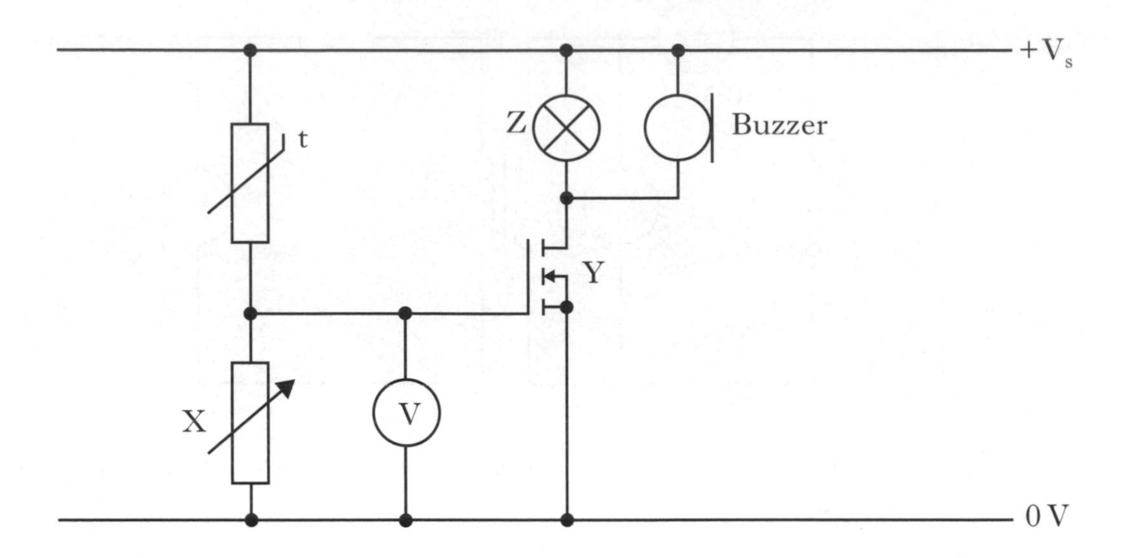

(*a*) Name components Y and Z. **2**

(*b*) What happens to the resistance of the thermistor as the temperature increases? **1**

(*c*) When the voltmeter reading reaches 1·8 V component Y switches on. Explain how the circuit operates when the temperature rises. **2**

(*d*) Why is a variable resistor chosen for component X rather than a fixed value resistor? **1**

(6)

Marks

27. At the kick-off in a football match, during the World Cup Finals, the referee blows his whistle. The whistle produces sound waves.

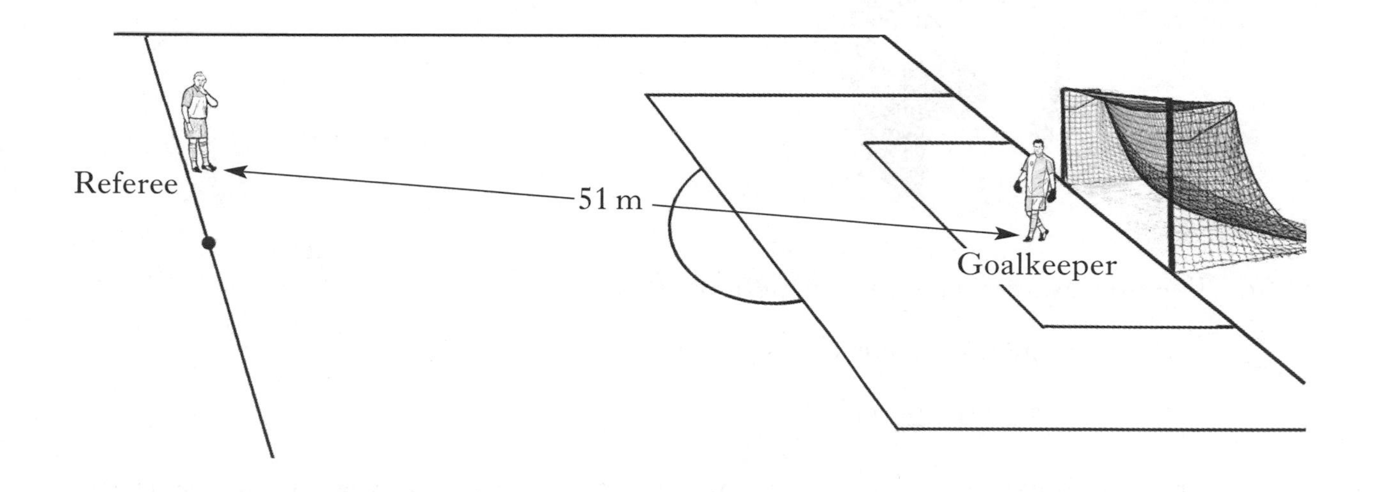

(*a*) Using information from the diagram and the data sheet, calculate the time taken for the sound waves to reach the goalkeeper. **2**

(*b*) (i) Are sound waves transverse or longitudinal waves? **1**

(ii) Describe **two** differences between transverse and longitudinal waves. **2**

(iii) What is transferred by waves? **1**

(*c*) (i) Floodlights in the stadium are switched on. Each lamp has a power rating of 2·40 kW. The operating voltage is 315 V.

Calculate the resistance of a lamp. **2**

(ii) The floodlights consist of 20 lamps connected in parallel.

State **two** reasons why the lamps are connected in parallel. **2**

(10)

[Turn over

Marks

28. A satellite sends microwaves to a ground station on Earth.

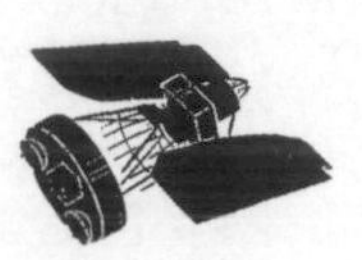

(*a*) The microwaves have a wavelength of 60 mm.

(i) Calculate the frequency of the waves. **2**

(ii) Determine the period of the waves. **2**

(*b*) The satellite sends radio waves along with the microwaves to the ground station. Will the radio waves be received by the ground station **before**, **after** or **at the same time** as the microwaves?

Explain your answer. **2**

(*c*) When the microwaves reach the ground station they are received by a curved reflector.

Explain why a curved reflector is used.

Your answer may include a diagram. **2**

(8)

Marks

29. In 1908 Ernest Rutherford conducted a series of experiments involving alpha particles.

(*a*) State what is meant by an alpha particle. **1**

(*b*) Alpha particles produce a greater ionisation density than beta particles or gamma rays. What is meant by the term *ionisation*? **1**

(*c*) A radioactive source emits alpha particles and has a half-life of 2·5 hours. The source has an initial activity of 4·8 kBq.

Calculate the time taken for its activity to decrease to 300 Bq. **2**

(*d*) Calculate the number of decays in the sample in two minutes, when the activity of the source is 1·2 kBq. **2**

(*e*) Some sources emit alpha particles and are stored in lead cases despite the fact that alpha particles cannot penetrate paper. Suggest a possible reason for storing these sources using this method. **1**

(7)

[Turn over for Question 30 on *Page twenty*

Marks

30. Many countries use nuclear reactors to produce energy. A diagram of the core of a nuclear reactor is shown.

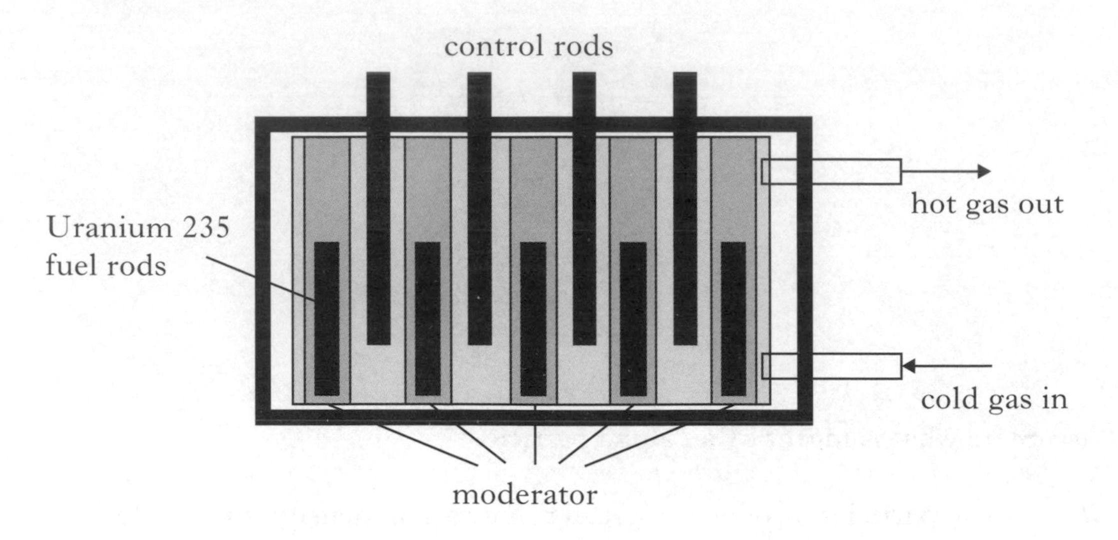

(*a*) State the purpose of:

(i) the moderator; **1**

(ii) the control rods. **1**

(*b*) One nuclear fission reaction produces $2{\cdot}9 \times 10^{-11}$ J of energy. The power output of the reactor is 1·4 GW. How many fission reactions are produced in one hour? **3**

(*c*) State **one advantage** and **one disadvantage** of using nuclear power for the generation of electricity. **2**

(7)

[*END OF QUESTION PAPER*]

INTERMEDIATE 2

2011

[BLANK PAGE]

X069/201

NATIONAL QUALIFICATIONS 2011

MONDAY, 23 MAY
1.00 PM – 3.00 PM

PHYSICS
INTERMEDIATE 2

Read Carefully

Reference may be made to the Physics Data Booklet

1 All questions should be attempted.

Section A (questions 1 to 20)

2 Check that the answer sheet is for Physics Intermediate 2 (Section A).

3 For this section of the examination you must use an **HB pencil** and, where necessary, an eraser.

4 Check that the answer sheet you have been given has **your name**, **date of birth**, **SCN** (Scottish Candidate Number) and **Centre Name** printed on it.

Do not change any of these details.

5 If any of this information is wrong, tell the Invigilator immediately.

6 If this information is correct, **print** your name and seat number in the boxes provided.

7 There is **only one correct** answer to each question.

8 Any rough working should be done on the question paper or the rough working sheet, **not** on your answer sheet.

9 At the end of the exam, **put the answer sheet for Section A inside the front cover of your answer book.**

10 Instructions as to how to record your answers to questions 1–20 are given on page three.

Section B (questions 21 to 31)

11 Answer the questions numbered 21 to 31 in the answer book provided.

12 **All answers must be written clearly and legibly in ink.**

13 Fill in the details on the front of the answer book.

14 Enter the question number clearly in the margin of the answer book beside each of your answers to questions 21 to 31.

15 Care should be taken to give an appropriate number of significant figures in the final answers to calculations.

LI X069/201 6/7310

DATA SHEET

Speed of light in materials

Material	*Speed* in m/s
Air	3.0×10^8
Carbon dioxide	3.0×10^8
Diamond	1.2×10^8
Glass	2.0×10^8
Glycerol	2.1×10^8
Water	2.3×10^8

Speed of sound in materials

Material	*Speed* in m/s
Aluminium	5200
Air	340
Bone	4100
Carbon dioxide	270
Glycerol	1900
Muscle	1600
Steel	5200
Tissue	1500
Water	1500

Gravitational field strengths

	Gravitational field strength on the surface in N/kg
Earth	10
Jupiter	26
Mars	4
Mercury	4
Moon	1·6
Neptune	12
Saturn	11
Sun	270
Venus	9

Specific heat capacity of materials

Material	*Specific heat capacity* in J/kg °C
Alcohol	2350
Aluminium	902
Copper	386
Glass	500
Ice	2100
Iron	480
Lead	128
Oil	2130
Water	4180

Specific latent heat of fusion of materials

Material	*Specific latent heat of fusion* in J/kg
Alcohol	0.99×10^5
Aluminium	3.95×10^5
Carbon Dioxide	1.80×10^5
Copper	2.05×10^5
Iron	2.67×10^5
Lead	0.25×10^5
Water	3.34×10^5

Melting and boiling points of materials

Material	*Melting point in °C*	*Boiling point in °C*
Alcohol	−98	65
Aluminium	660	2470
Copper	1077	2567
Glycerol	18	290
Lead	328	1737
Iron	1537	2737

Specific latent heat of vaporisation of materials

Material	*Specific latent heat of vaporisation* in J/kg
Alcohol	11.2×10^5
Carbon Dioxide	3.77×10^5
Glycerol	8.30×10^5
Turpentine	2.90×10^5
Water	22.6×10^5

Radiation weighting factors

Type of radiation	*Radiation weighting factor*
alpha	20
beta	1
fast neutrons	10
gamma	1
slow neutrons	3

SECTION A

For questions 1 to 20 in this section of the paper the answer to each question is either A, B, C, D or E. Decide what your answer is, then, using your pencil, put a horizontal line in the space provided—see the example below.

EXAMPLE

The energy unit measured by the electricity meter in your home is the

A kilowatt-hour

B ampere

C watt

D coulomb

E volt.

The correct answer is **A**—kilowatt-hour. The answer **A** has been clearly marked in **pencil** with a horizontal line (see below).

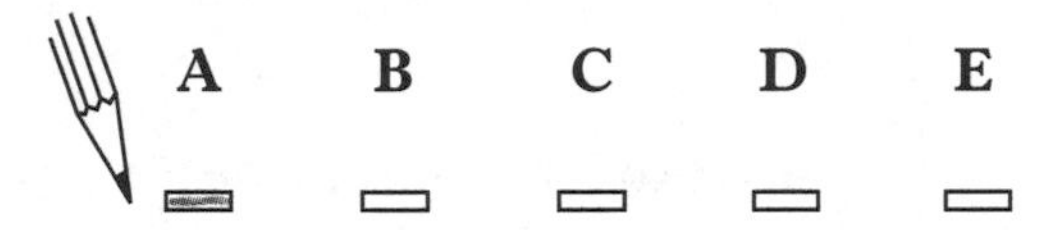

Changing an answer

If you decide to change your answer, carefully erase your first answer and, using your pencil, fill in the answer you want. The answer below has been changed to **E**.

A B C D E

[Turn over

SECTION A

Answer questions 1–20 on the answer sheet.

1. During training an athlete sprints 30 m East and then 40 m West.

Which row shows the distance travelled and the displacement from the starting point?

	Distance travelled	*Displacement*
A	10 m	10 m East
B	10 m	10 m West
C	10 m	70 m East
D	70 m	10 m West
E	70 m	10 m East

2. The graph shows how the velocity of a ball changes with time.

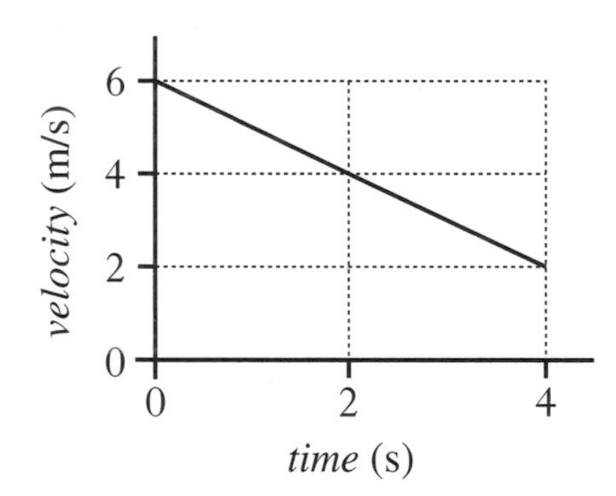

The acceleration of the ball is

A $-8\ m/s^2$

B $-1\ m/s^2$

C $1\ m/s^2$

D $8\ m/s^2$

E $24\ m/s^2$.

3. A ball of mass 2 kg moves along a horizontal surface at 4 m/s.

Which row shows the momentum and kinetic energy of the ball?

	Momentum (kg m/s)	*Kinetic energy* (J)
A	2	4
B	4	8
C	4	16
D	8	8
E	8	16

4. An engine applies a force of 2000 N to move a lorry at a constant speed.

The lorry travels 100 m in 16 s.

The power developed by the engine is

A 0·8 W

B 12·5 W

C 320 W

D 12 500 W

E 3 200 000 W.

5. Which row in the table identifies the following circuit symbols?

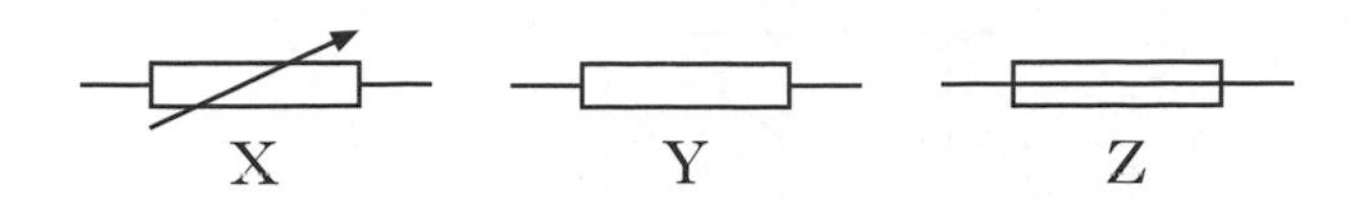

	Symbol X	*Symbol Y*	*Symbol Z*
A	fuse	resistor	variable resistor
B	fuse	variable resistor	resistor
C	resistor	variable resistor	fuse
D	variable resistor	fuse	resistor
E	variable resistor	resistor	fuse

6. Which graph shows how the potential difference V across a resistor varies with the current I in the resistor?

A 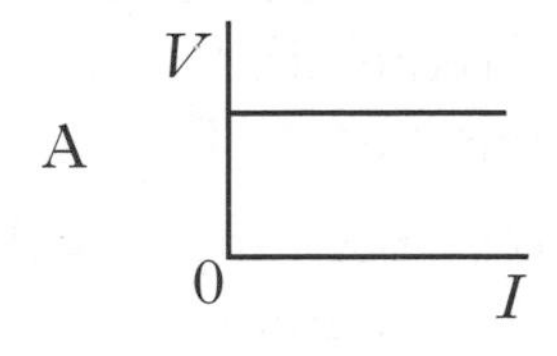

B 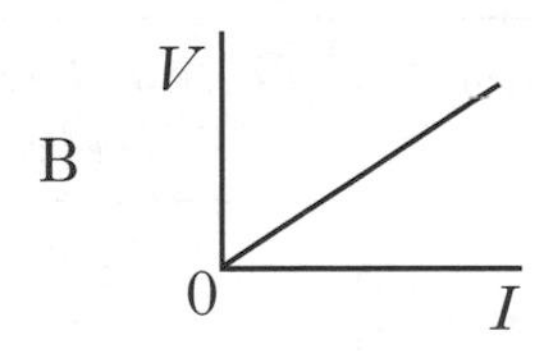

C 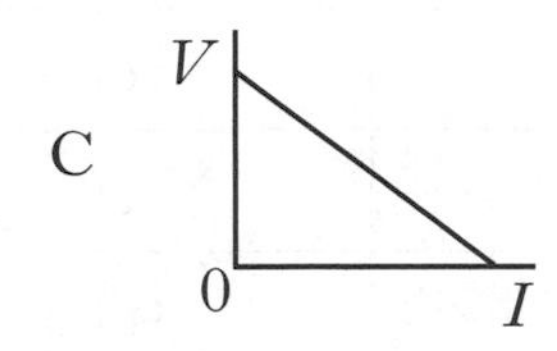

D 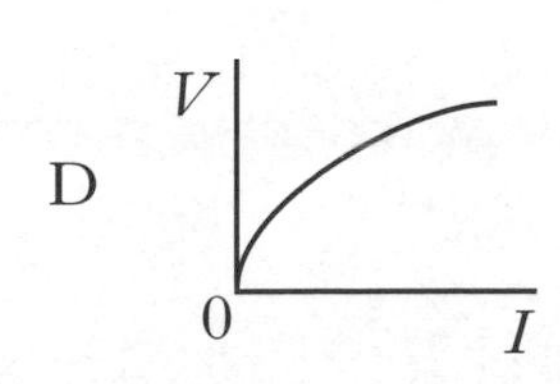

E 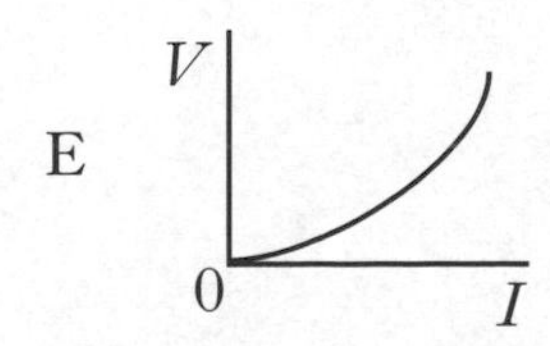

7. A circuit is set up as shown.

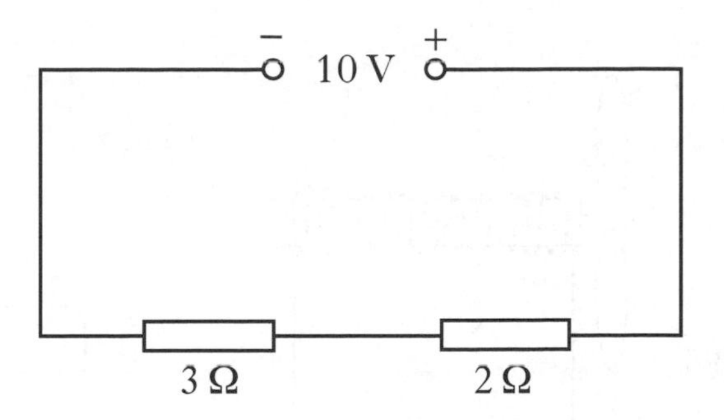

The potential difference across the 2 Ω resistor is

A 4 V

B 5 V

C 6 V

D 10 V

E 20 V.

8. A student makes the following statements about electrical supplies.

I The frequency of the mains supply is 50 Hz.

II The quoted value of an alternating voltage is less than its peak value.

III A d.c. supply and an a.c. supply of the same quoted value will supply the same power to a given resistor.

Which of the following statements is/are correct?

A I only

B II only

C III only

D I and II only

E I, II and III

[Turn over

9. A wind speed meter is designed as shown.

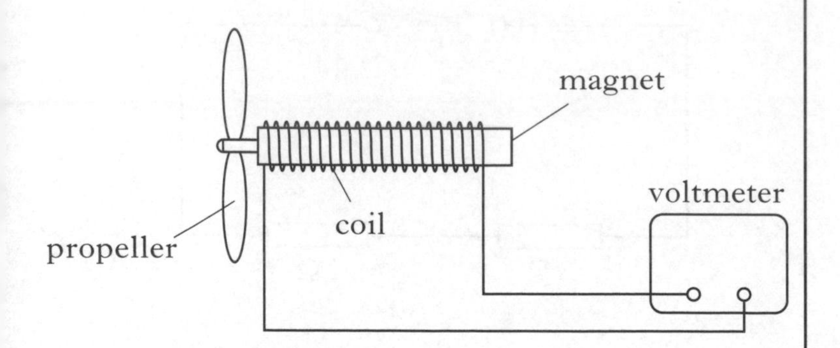

Air blows across the propeller causing the magnet to rotate. A voltage is induced across the coil.

Which of the following changes will produce an increase in the induced voltage?

I Replacing the magnet with one of greater field strength.

II Spinning the propeller faster.

III Reducing the number of turns on the coil.

A I only

B I and II only

C I and III only

D II and III only

E I, II and III

10. Which of the following is the circuit symbol for an NPN transistor?

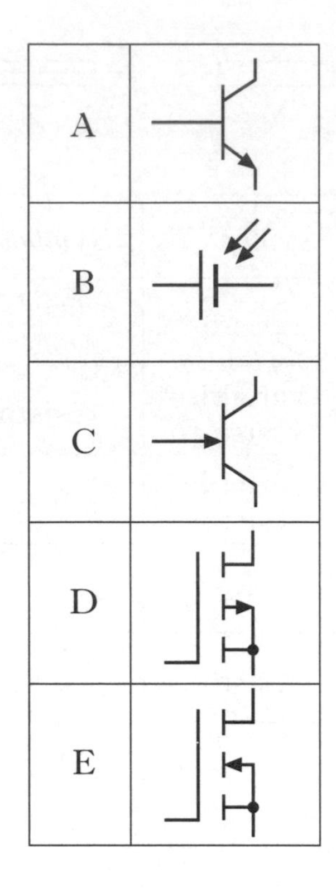

11. The input signal to an amplifier is 2 V a.c. at a frequency of 200 Hz. The amplifier has a gain of 8.

Which row shows the output voltage and the output frequency?

	Output voltage (V)	*Output frequency* (Hz)
A	10	200
B	10	208
C	10	1600
D	16	200
E	16	1600

12. The following diagram gives information about a wave.

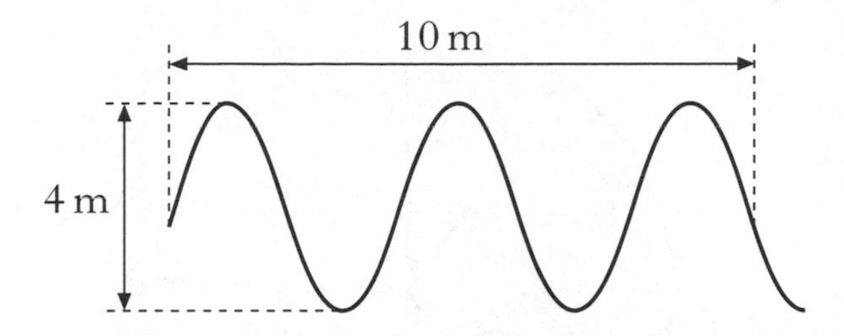

Which row shows the amplitude and wavelength of the wave?

	Amplitude (m)	*Wavelength* (m)
A	2	2
B	2	4
C	2	5
D	4	2
E	4	4

13. Sound is a longitudinal wave. When a sound wave travels through air the particles of air

A vibrate at random

B vibrate along the wave direction

C vibrate at 90 ° to the wave direction

D move continuously away from the source

E move continuously towards the source.

14. A signal is transmitted using a curved reflector as shown.

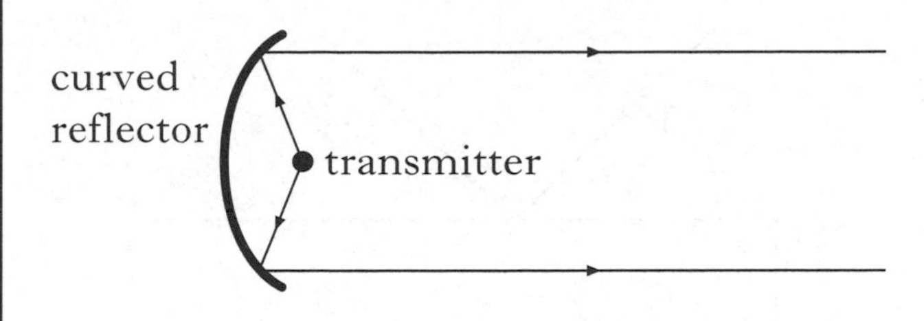

Which of the following statements is/are correct?

I The signal meets the curved reflector at an angle called the critical angle.

II The transmitter is placed at the focus of the reflector.

III At the curved reflector, the angle of reflection of the signal is equal to the angle of incidence.

A I only

B I and II only

C I and III only

D II and III only

E I, II and III

[Turn over

15. The diagram shows a ray of light P incident on a rectangular glass block.

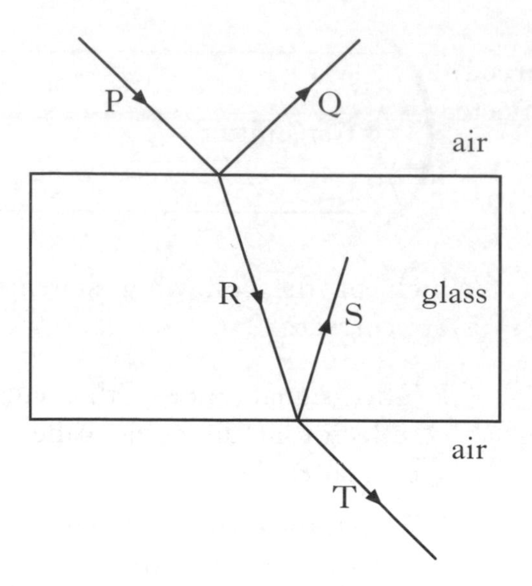

Which of the following are refracted rays?

A Q and R

B R and S

C S and T

D Q and S

E R and T

16. The diagram shows the path of a ray of red light in a glass block.

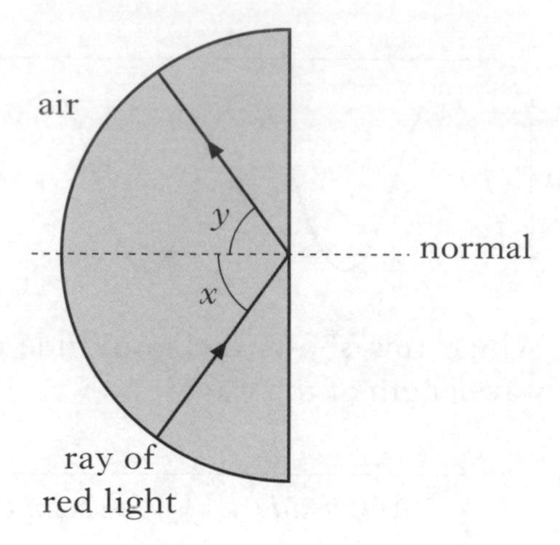

A student makes the following statements.

I Angle x is equal to angle y.

II Total internal reflection is taking place.

III Angle x is the critical angle for this glass.

Which of the following statements is/are correct?

A I only

B II only

C I and II only

D II and III only

E I, II and III

17. Activity and absorbed dose are quantities used in Dosimetry.

Which row shows the unit of activity and the unit of absorbed dose?

	Unit of activity	*Unit of absorbed dose*
A	gray	becquerel
B	becquerel	sievert
C	becquerel	gray
D	gray	sievert
E	sievert	gray

18. The table shows the count rate of a radioactive source taken at regular time intervals. The count rate has been corrected for background radiation.

ime (minutes)	10	20	30	40	50
ount rate (ounts per minute)	800	630	500	400	315

What is the half-life in minutes of the isotope?

A 10

B 15

C 20

D 30

E 40

19. In the following passage some words have been replaced by letters X and Y.

In a nuclear reactor, fission is caused by X bombardment of a uranium nucleus. This causes the nucleus to split releasing neutrons and Y.

Which row gives the words for X and Y?

	X	*Y*
A	neutron	energy
B	proton	energy
C	electron	protons
D	neutron	protons
E	electron	energy

20. Control rods in a nuclear reactor

A absorb neutrons

B contain uranium

C produce neutrons

D remove heat from the reactor

E slow down neutrons.

Candidates are reminded that the answer sheet for Section A MUST be placed INSIDE the front cover of the answer book.

[Turn over

SECTION B

Marks

Write your answers to questions 21–31 in the answer book.

All answers must be written clearly and legibly in ink.

21. A cricketer strikes a ball. The ball leaves the bat horizontally at 20 m/s. It hits the ground at a distance of 11 m from the point where it was struck.

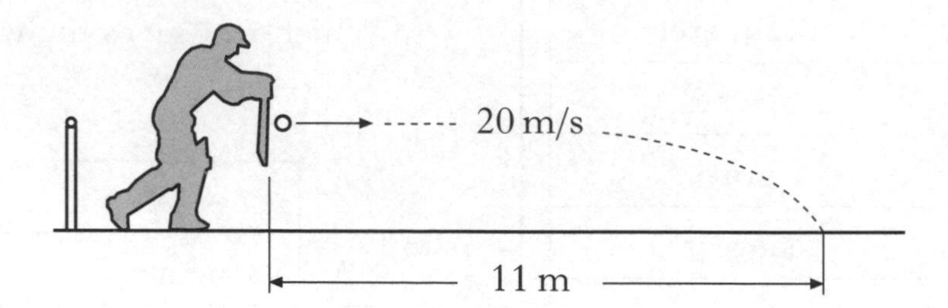

Assume that air resistance is negligible.

(*a*) Calculate the time of flight of the ball. **2**

(*b*) Calculate the vertical speed of the ball as it reaches the ground. **2**

(*c*) Sketch a graph of vertical speed against time for the ball. Numerical values are required on both axes. **2**

(*d*) Calculate the vertical distance travelled by the ball during its flight. **2**

(8)

Marks

22. A satellite moves in a circular orbit around a planet. The satellite travels at a constant speed whilst accelerating.

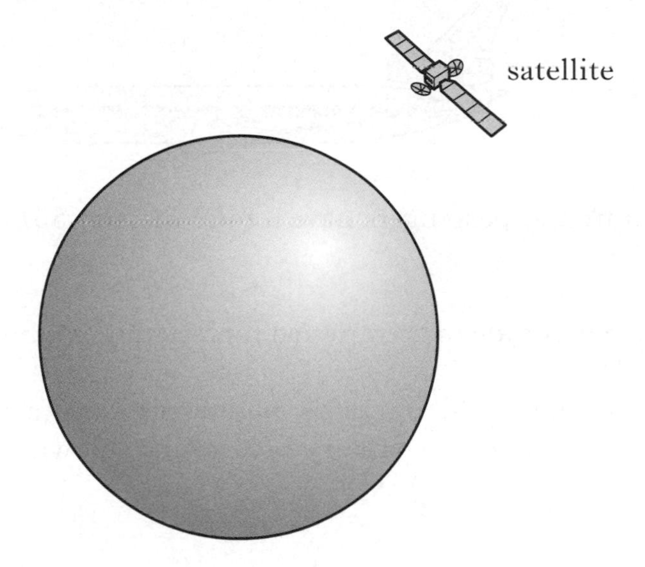

(*a*) (i) Define the term *acceleration*. **1**

(ii) Explain how the satellite can be accelerating when it is travelling at a constant speed. **1**

(*b*) At one particular point in its orbit the satellite fires two rockets. The forces exerted on the satellite by these rockets are shown on the diagram.

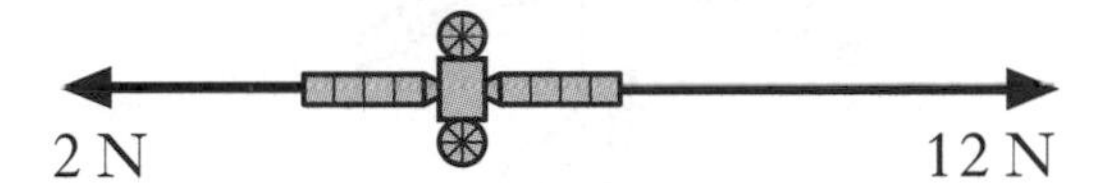

The satellite has a mass of 50 kg. Calculate the resultant acceleration due to these forces. **3**

(5)

[Turn over

Marks

23. An aircraft is flying horizontally at a constant speed.

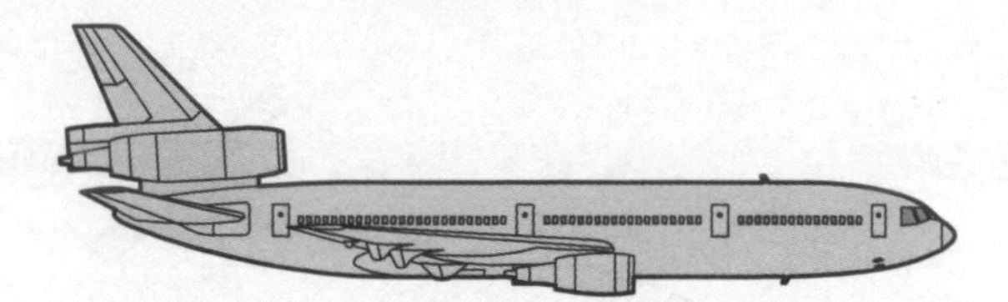

(*a*) The aircraft and passengers have a total mass of 50 000 kg. Calculate the total weight. **2**

(*b*) State the magnitude of the upward force acting on the aircraft. **1**

(*c*) During the flight, the aircraft's engines produce a force of $4{\cdot}4 \times 10^4$ N due North. The aircraft encounters a crosswind, blowing from west to east, which exerts a force of $3{\cdot}2 \times 10^4$ N.

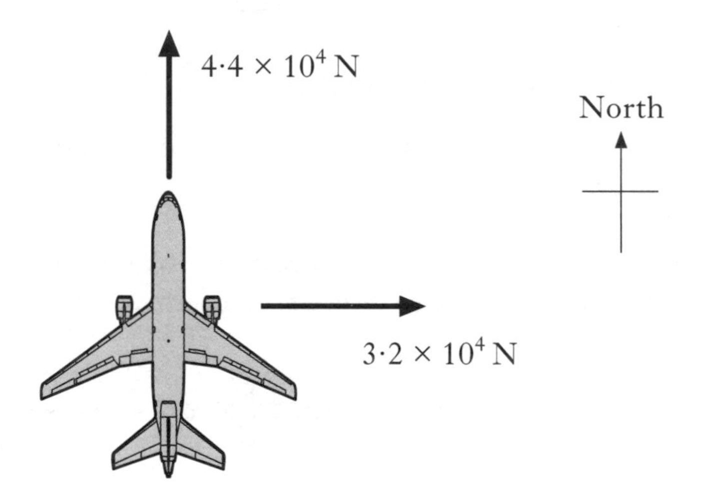

Calculate the resultant force on the aircraft. **3**

(*d*) During a particular flight, a pilot receives an absorbed dose of 15 μGy from gamma rays. Calculate the equivalent dose received due to this type of radiation. **2**

(*e*) Gamma radiation is an example of radiation which causes *ionisation*. Explain what is meant by the term *ionisation*. **1**

(9)

Marks

24. An experiment was carried out to determine the specific heat capacity of water. The energy supplied to the water was measured by a joulemeter.

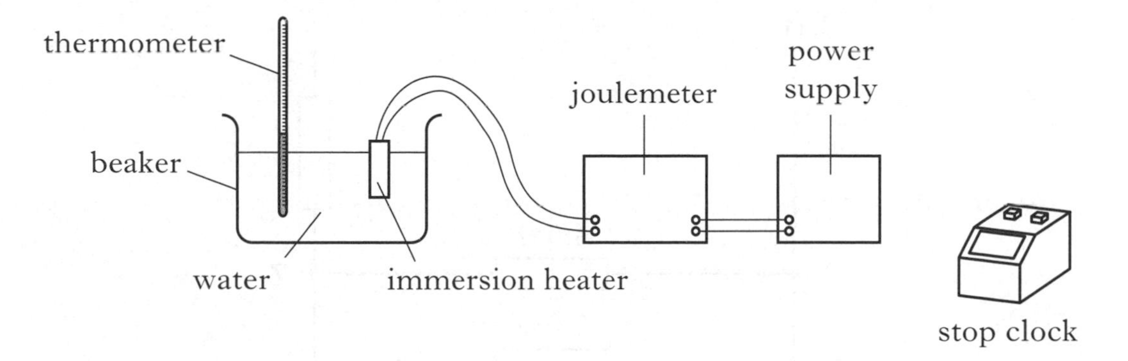

The following data was recorded.

Initial temperature of the water = 21 °C.
Final temperature of the water = 33 °C.
Initial reading on the joulemeter = 12 kJ.
Final reading on the joulemeter = 120 kJ.
Mass of water = 2·0 kg.
Time = 5 minutes.

(*a*) (i) Calculate the change in temperature of the water. **1**

(ii) Calculate the energy supplied by the immersion heater. **1**

(iii) Calculate the value for the specific heat capacity of water obtained from this experiment. **2**

(*b*) (i) The accepted value for the specific heat capacity of water is quoted in the table in the Data Sheet. Explain the difference between the accepted value and the value obtained in the experiment. **2**

(ii) How could the experiment be improved to reduce this difference? **1**

(*c*) Calculate the power rating of the immersion heater. **2**

(9)

[Turn over

Marks

25. Part of a circuit is shown below.

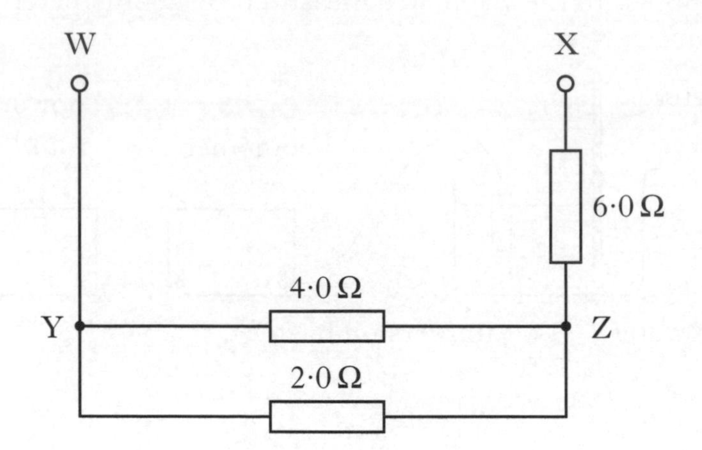

(*a*) Calculate the total resistance between points Y and Z. **2**

(*b*) Calculate the total resistance between points W and X. **2**

(*c*) Calculate the voltage across the 2·0 Ω resistor when the current in the 4·0 Ω resistor is 0·10 A. **2**

(6)

Marks

26. A student has two electrical power supplies. One is an a.c. supply and the other is a d.c. supply.

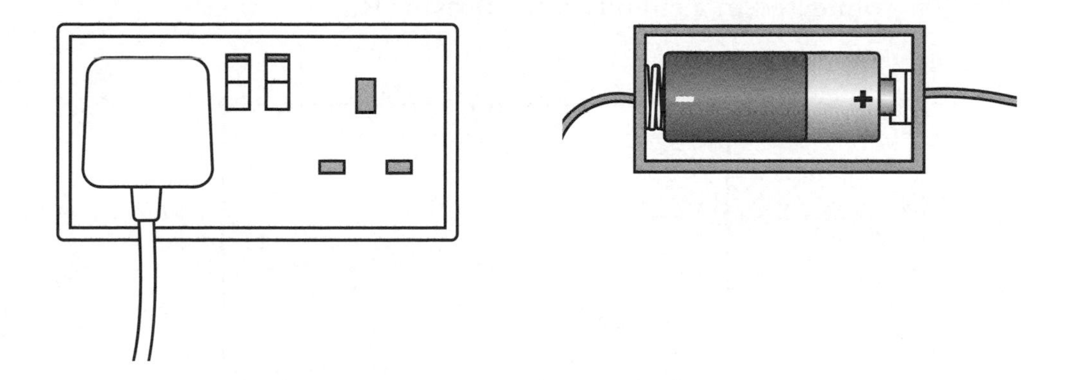

(*a*) Explain a.c. and d.c. in terms of electron flow in a circuit. **2**

(*b*) The student uses **one** of the supplies to operate a transformer.

(i) Which power supply should be used to operate the transformer? **1**

(ii) What is the purpose of a transformer? **1**

(iii) A transformer with an efficiency of 30% is used in a computer. Calculate the output power when the input power is 50 W. **2**

(6)

[Turn over

Marks

27. Light emitting diodes (LEDs) are often used as on/off indicators on televisions and computers.

An LED is connected in a circuit with a resistor R.

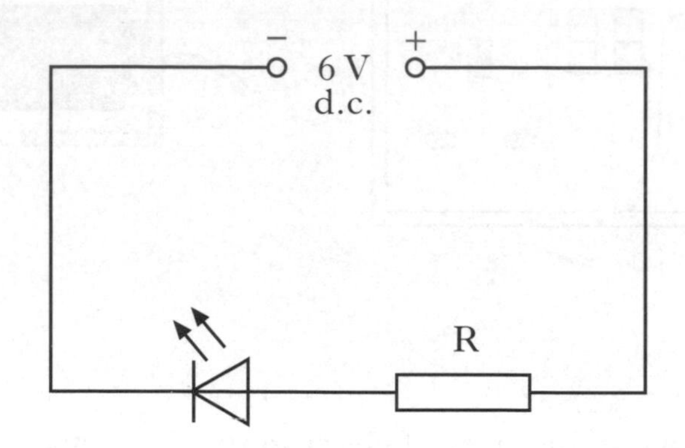

(*a*) What is the purpose of resistor R? **1**

(*b*) The LED is rated at 2 V, 100 mA. Calculate the resistance of resistor R. **3**

(*c*) Calculate the power developed by resistor R when the LED is working normally. **2**

(6)

Marks

28. A solar cell is tested for use in a buggy.

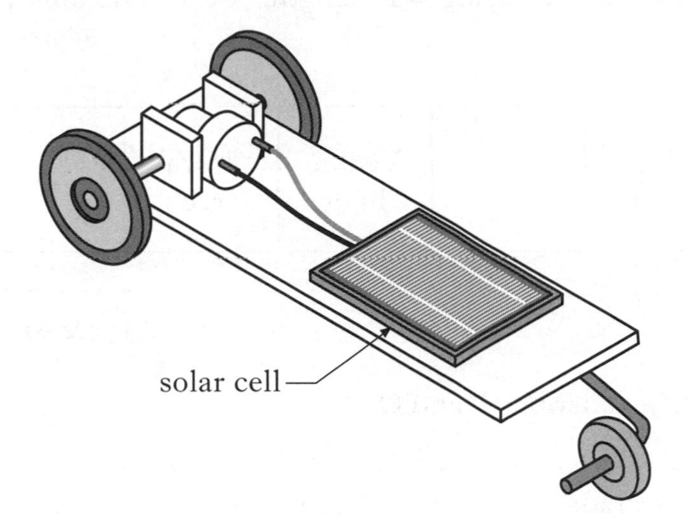

The solar cell produces a voltage of 0·5 V and a current of 0·4 mA.

(*a*) (i) Calculate the power produced by the solar cell. **2**

(ii) The buggy requires 4 mW to operate. Calculate the number of solar cells required to supply this power. **2**

(*b*) State the energy change in a solar cell. **1**

(*c*) The solar cell is illuminated by light of frequency $6{\cdot}7 \times 10^{14}$ Hz. Calculate the wavelength of this light. **2**

(7)

[Turn over

Marks

29. The Sun produces electromagnetic radiation. The electromagnetic spectrum is shown in order of increasing wavelength. Two radiations P and Q have been omitted.

Gamma rays	X rays	P	Visible light	Infra red	Q	Television and radio rays

Increasing wavelength →

(*a*) (i) Identify radiations P and Q. **2**

(ii) The planet Neptune is $4 \cdot 50 \times 10^9$ km from the Sun. Calculate the time taken for radio waves from the Sun to reach Neptune. **2**

(iii) State what happens to the frequency of electromagnetic radiation as the wavelength increases. **1**

(*b*) The Sun produces a *solar wind* consisting of charged particles. In one particular part of the solar wind, a charge of 360 C passes a point in space in one minute. Calculate the current. **2**

(7)

Marks

30. A converging lens has a focal length of 30 mm.

(*a*) Calculate the power of the lens. **2**

(*b*) On the graph paper provided, copy and complete the diagram below.
Show the size and position of the image formed by the lens. **3**

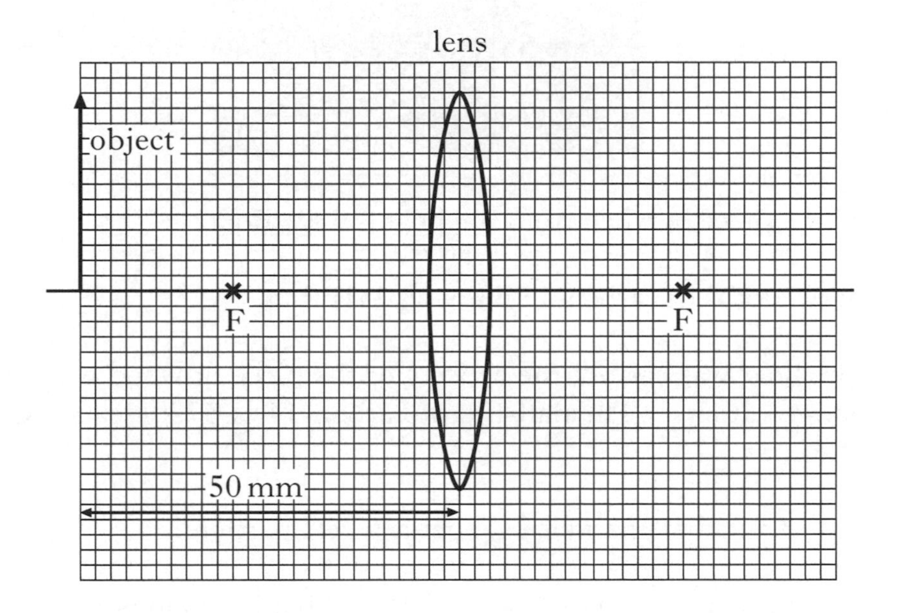

(*c*) Name the eye defect which a converging lens can correct. Explain your answer. **2**

(7)

[Turn over for Question 31 on *Page twenty*

Marks

31. It is possible to determine the age of a prehistoric wooden boat by measuring the activity of radioactive carbon-14.

The activity of a piece of wood from the boat is 300 μBq.

(*a*) How many atoms of carbon-14 decay in 1 day? **2**

(*b*) When the boat was carved, the activity of the piece of wood was 2400 μBq due to carbon-14 atoms. The half-life of carbon-14 is 5730 years. Calculate the age of the boat. **2**

(*c*) Carbon-14 emits beta particles. What is a beta particle? **1**

(*d*) A radioactive source emits alpha particles. What is an alpha particle? **1**

(*e*) How does the ionisation density of alpha particles compare with that of beta particles? **1**

(*f*) (i) A student sets up an experiment as shown.

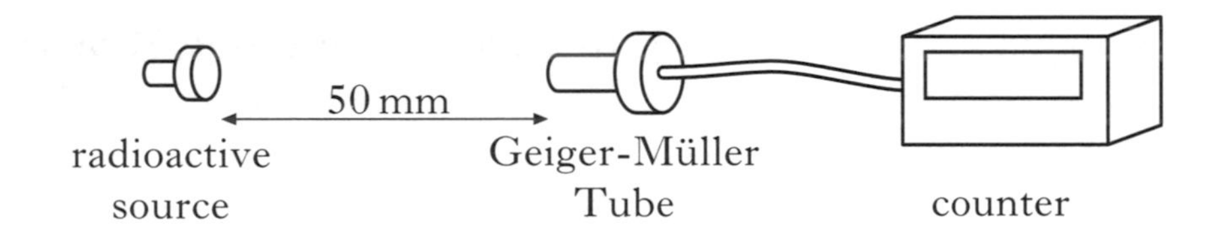

The student places a 3 mm sheet of aluminium between the radioactive source and the Geiger-Müller Tube. The count rate is observed to decrease and the student concludes that the radioactive material is emitting beta radiation.

Suggest **one** reason why her conclusion may be incorrect. **1**

(ii) State **two** safety precautions that the student must observe when handling radioactive sources. **2**

(10)

[*END OF QUESTION PAPER*]

INTERMEDIATE 2

2012

[BLANK PAGE]

X069/11/02

NATIONAL QUALIFICATIONS 2012

MONDAY, 28 MAY
1.00 PM – 3.00 PM

PHYSICS
INTERMEDIATE 2

Read Carefully

Reference may be made to the Physics Data Booklet

1 All questions should be attempted.

Section A (questions 1 to 20)

2 Check that the answer sheet is for Physics Intermediate 2 (Section A).

3 For this section of the examination you must use an **HB pencil** and, where necessary, an eraser.

4 Check that the answer sheet you have been given has **your name**, **date of birth**, **SCN** (Scottish Candidate Number) and **Centre Name** printed on it.

Do not change any of these details.

5 If any of this information is wrong, tell the Invigilator immediately.

6 If this information is correct, **print** your name and seat number in the boxes provided.

7 There is **only one correct** answer to each question.

8 Any rough working should be done on the question paper or the rough working sheet, **not** on your answer sheet.

9 At the end of the exam, **put the answer sheet for Section A inside the front cover of your answer book**.

10 Instructions as to how to record your answers to questions 1–20 are given on page three.

Section B (questions 21 to 30)

11 Answer the questions numbered 21 to 30 in the answer book provided.

12 **All answers must be written clearly and legibly in ink.**

13 Fill in the details on the front of the answer book.

14 Enter the question number clearly in the margin of the answer book beside each of your answers to questions 21 to 30.

15 Care should be taken to give an appropriate number of significant figures in the final answers to calculations.

LI X069/11/02 6/7710

DATA SHEET

Speed of light in materials

Material	*Speed* in m/s
Air	$3{\cdot}0 \times 10^8$
Carbon dioxide	$3{\cdot}0 \times 10^8$
Diamond	$1{\cdot}2 \times 10^8$
Glass	$2{\cdot}0 \times 10^8$
Glycerol	$2{\cdot}1 \times 10^8$
Water	$2{\cdot}3 \times 10^8$

Speed of sound in materials

Material	*Speed* in m/s
Aluminium	5200
Air	340
Bone	4100
Carbon dioxide	270
Glycerol	1900
Muscle	1600
Steel	5200
Tissue	1500
Water	1500

Gravitational field strengths

	Gravitational field strength on the surface in N/kg
Earth	10
Jupiter	26
Mars	4
Mercury	4
Moon	1·6
Neptune	12
Saturn	11
Sun	270
Venus	9

Specific heat capacity of materials

Material	*Specific heat capacity* in J/kg °C
Alcohol	2350
Aluminium	902
Copper	386
Glass	500
Ice	2100
Iron	480
Lead	128
Oil	2130
Water	4180

Specific latent heat of fusion of materials

Material	*Specific latent heat of fusion* in J/kg
Alcohol	$0{\cdot}99 \times 10^5$
Aluminium	$3{\cdot}95 \times 10^5$
Carbon Dioxide	$1{\cdot}80 \times 10^5$
Copper	$2{\cdot}05 \times 10^5$
Iron	$2{\cdot}67 \times 10^5$
Lead	$0{\cdot}25 \times 10^5$
Water	$3{\cdot}34 \times 10^5$

Melting and boiling points of materials

Material	*Melting point in °C*	*Boiling point in °C*
Alcohol	−98	65
Aluminium	660	2470
Copper	1077	2567
Glycerol	18	290
Lead	328	1737
Iron	1537	2737

Specific latent heat of vaporisation of materials

Material	*Specific latent heat of vaporisation* in J/kg
Alcohol	$11{\cdot}2 \times 10^5$
Carbon Dioxide	$3{\cdot}77 \times 10^5$
Glycerol	$8{\cdot}30 \times 10^5$
Turpentine	$2{\cdot}90 \times 10^5$
Water	$22{\cdot}6 \times 10^5$

Radiation weighting factors

Type of radiation	*Radiation weighting factor*
alpha	20
beta	1
fast neutrons	10
gamma	1
slow neutrons	3

SECTION A

For questions 1 to 20 in this section of the paper the answer to each question is either A, B, C, D or E. Decide what your answer is, then, using your pencil, put a horizontal line in the space provided—see the example below.

EXAMPLE

The energy unit measured by the electricity meter in your home is the

A kilowatt-hour

B ampere

C watt

D coulomb

E volt.

The correct answer is **A**—kilowatt-hour. The answer **A** has been clearly marked in **pencil** with a horizontal line (see below).

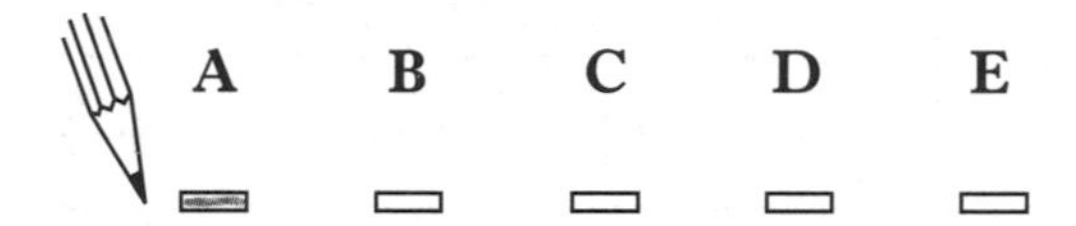

Changing an answer

If you decide to change your answer, carefully erase your first answer and, using your pencil, fill in the answer you want. The answer below has been changed to **E**.

A B C D E

[Turn over

SECTION A

Answer questions 1–20 on the answer sheet.

1. At an airport an aircraft moves from the terminal building to the end of the runway.

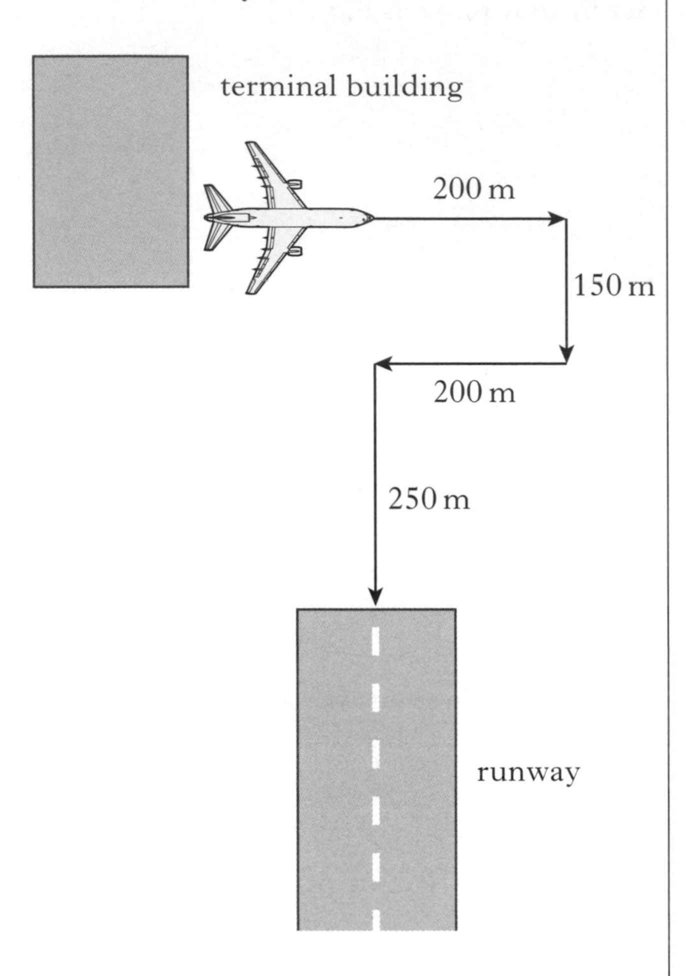

Which row shows the total distance travelled and the size of the displacement of the aircraft?

	Total distance travelled (m)	*Size of displacement* (m)
A	400	800
B	450	200
C	450	400
D	800	400
E	800	800

2. Near the Earth's surface, a mass of 6 kg is falling with a constant velocity.

The air resistance and the unbalanced force acting on the mass are:

	air resistance	*unbalanced force*
A	60 N upwards	0 N
B	10 N upwards	10 N downwards
C	10 N downwards	70 N downwards
D	10 N upwards	0 N
E	60 N upwards	60 N downwards

3. Two forces act on an object O in the directions shown.

The size of the resultant force is

A 14 N

B 24 N

C 38 N

D 45 N

E 62 N.

4. The diagram shows the horizontal forces acting on a box.

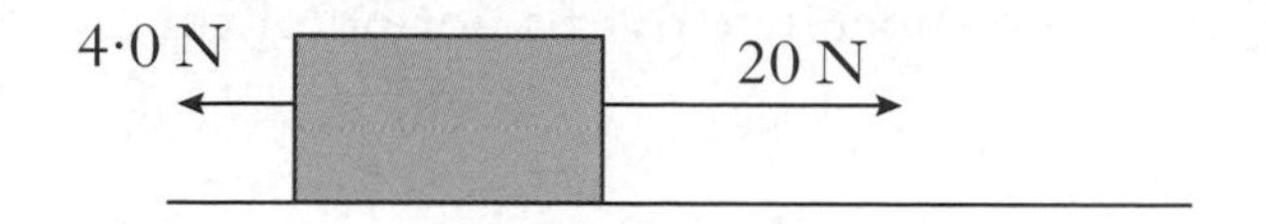

The box accelerates at 1·6 m/s^2.

The mass of the box is

A 0·10 kg

B 10·0 kg

C 15·0 kg

D 25·6 kg

E 38·4 kg.

5. Two identical balls X and Y are projected horizontally from the edge of a cliff.

The path taken by each ball is shown.

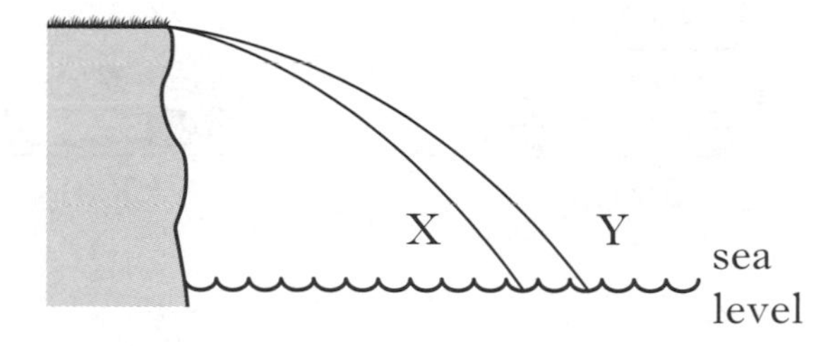

A student makes the following statements about the motion of the two balls.

I They take the same time to reach sea level.

II They have the same vertical acceleration.

III They have the same horizontal velocity.

Which of these statements is/are correct?

A I only

B II only

C I and II only

D I and III only

E II and III only

[Turn over

6. Car X of mass 1500 kg travels at 20 m/s along a straight, horizontal road. It collides with a stationary car Y of mass 1900 kg.

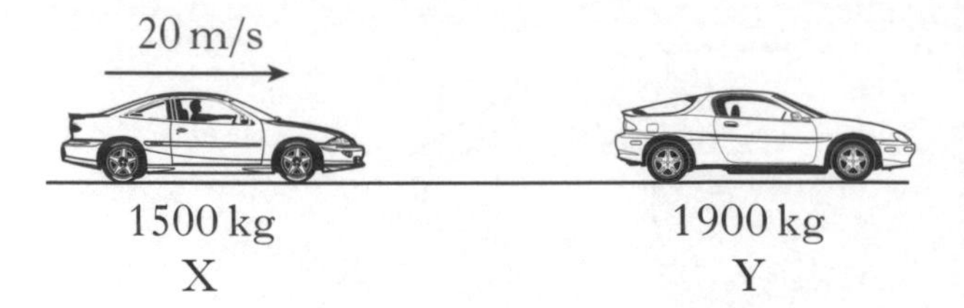

The two cars lock together after the collision.

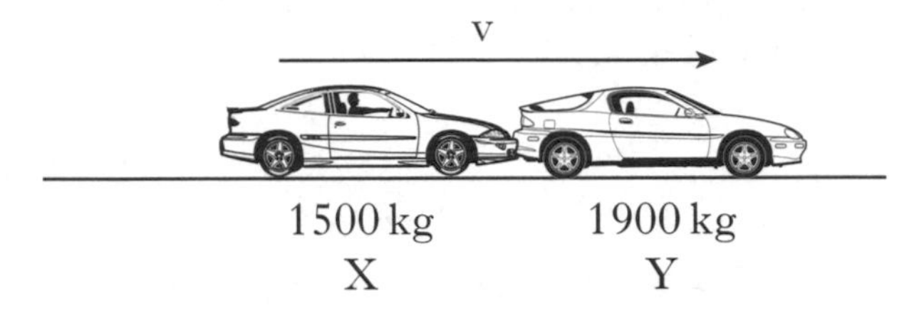

The speed of the cars after the collision is

A 8·8 m/s

B 9·4 m/s

C 11 m/s

D 16 m/s

E 20 m/s.

7. An electrical motor raises a crate of mass 500 kg through a height of 12 m in 4 s.

The minimum power rating of the motor is

A 1·25 kW

B 1·5 kW

C 15 kW

D 60 kW

E 240 kW.

8. A heater is immersed in a substance. The heater is then switched on.

The graph shows the temperature of the substance over a period of time.

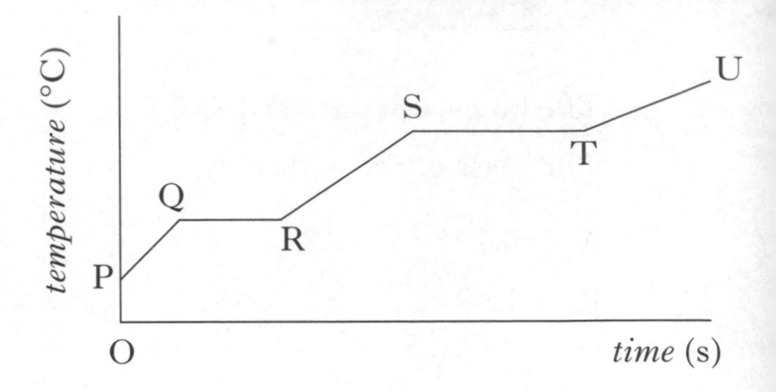

Which row in the table identifies the sections of the graph when the substance is changing state?

	Solid to liquid	*Liquid to gas*
A	QR	TU
B	QR	ST
C	PQ	RS
D	PQ	TU
E	ST	QR

9. Which row in the table gives the accepted values for the UK mains supply?

	Frequency (Hz)	*Quoted voltage* (V)	*Peak voltage* (V)
A	10	110	230
B	50	230	230
C	50	230	325
D	60	230	162
E	230	50	50

10. A circuit contains an ideal transformer connected to a 10 V d.c. supply.

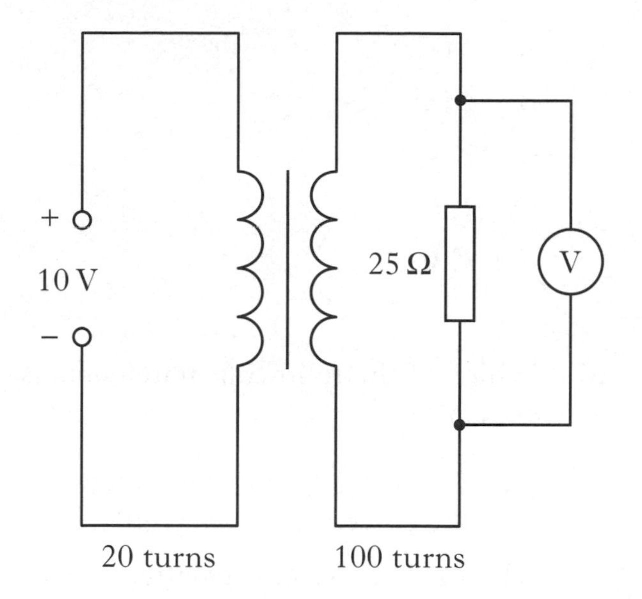

The potential difference across the 25 Ω resistor is

A 0 V

B 2 V

C 10 V

D 50 V

E 80 V.

11. A student sets up the circuits shown.

In which circuit will both LEDs be lit?

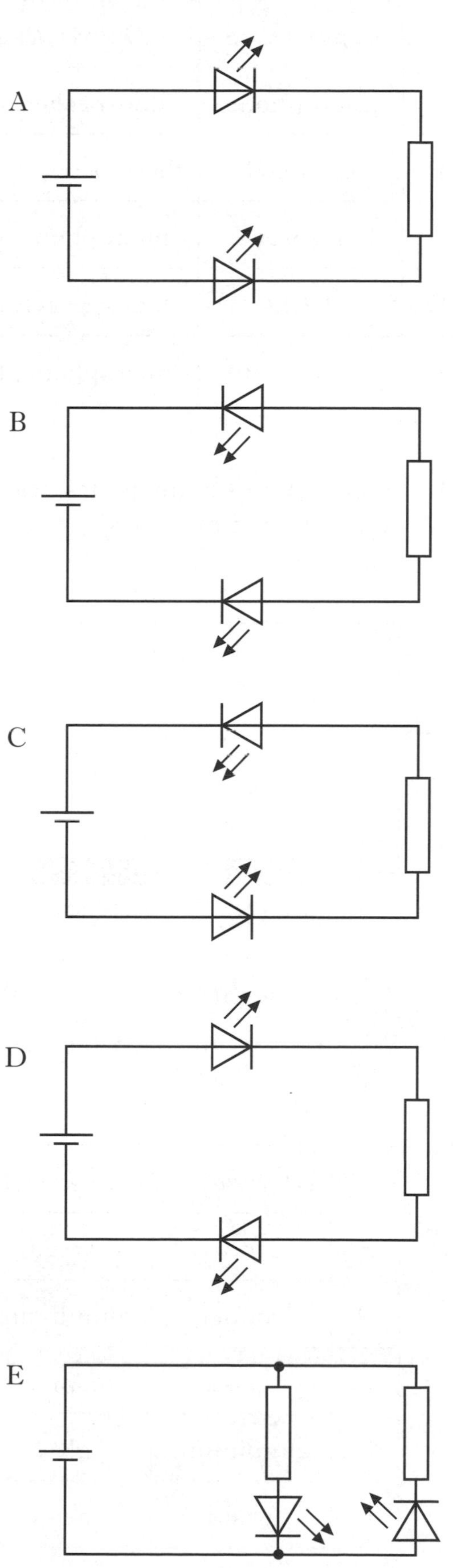

12. Which row in the table correctly identifies input and output devices?

	Input device	*Output devices*
A	microphone	loudspeaker, LED
B	solar cell	thermocouple, LED
C	loudspeaker	microphone, relay
D	LED	loudspeaker, relay
E	thermocouple	microphone, LED

13. A circuit is set up to test electrical conduction in materials.

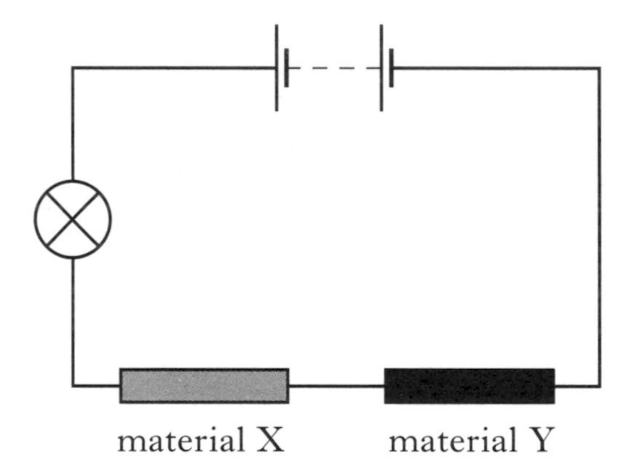

The lamp lights.

Which row in the table identifies materials X and Y?

	Material X	*Material Y*
A	copper	wood
B	copper	aluminium
C	glass	copper
D	aluminium	glass
E	wood	glass

14. The current in an 8 Ω resistor is 2 A.

The charge passing through the resistor in 10 s is

A 4 C

B 5 C

C 16 C

D 20 C

E 80 C.

15. Which of the following statements is/are correct?

I In an a.c. circuit the direction of the current changes regularly.

II In a d.c. circuit positive charges flow in one direction only.

III In an a.c. circuit the size of the current varies with time.

A I only

B II only

C I and II only

D I and III only

E I, II and III

16. A signal of voltage 5·0 mV and frequency 2000 Hz is applied to the input of an amplifier.

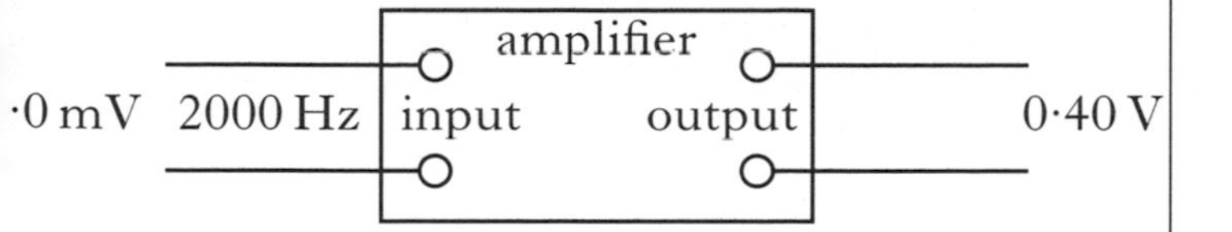

The output voltage is 0·40 V.

Which row shows the voltage gain of the amplifier and the frequency of the output signal?

	Voltage gain	*Frequency of output signal* (Hz)
A	0·0125	2000
B	0·08	50
C	0·08	2000
D	80	50
E	80	2000

17. The diagram shows two rays of red light X and Y passing through a block of glass.

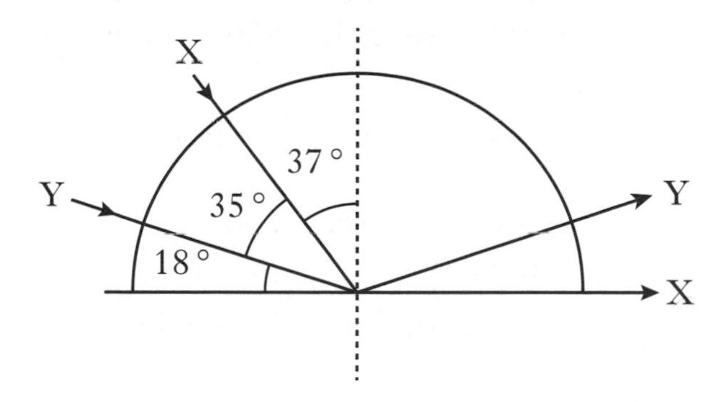

The critical angle of the glass for this light is

A 18 °

B 35 °

C 37 °

D 53 °

E 72 °.

18. A student makes the following statements.

I In an atom there are neutrons and electrons in the nucleus and protons which orbit the nucleus.

II An alpha particle consists of two neutrons and two electrons.

III A beta particle is a fast moving electron.

Which of the statements is/are correct?

A I only

B II only

C III only

D I and III only

E I, II and III

[Turn over

19. A radioactive source emits alpha, beta and gamma radiation. A detector, connected to a counter, is placed 10 mm in front of the source. The counter records 400 counts per minute.

A sheet of paper is placed between the source and the detector. The counter records 300 counts per minute.

The radiation now detected is

A alpha only

B beta only

C gamma only

D alpha and beta only

E beta and gamma only.

20. A radioactive tracer is injected into a patient to study the flow of blood.

The tracer should have a

A short half-life and emit α particles

B long half-life and emit β particles

C long half-life and emit γ rays

D long half-life and emit α particles

E short half-life and emit γ rays.

Candidates are reminded that the answer sheet for Section A MUST be placed INSIDE the front cover of the answer book.

[Turn over for Section B on *Page twelve*

SECTION B

Marks

Write your answers to questions 21–30 in the answer book.

All answers must be written clearly and legibly in ink.

21. Sputnik 1, the first man-made satellite, was launched in 1957. It orbited the Earth at a speed of 8300 m/s and had a mass of 84 kg.

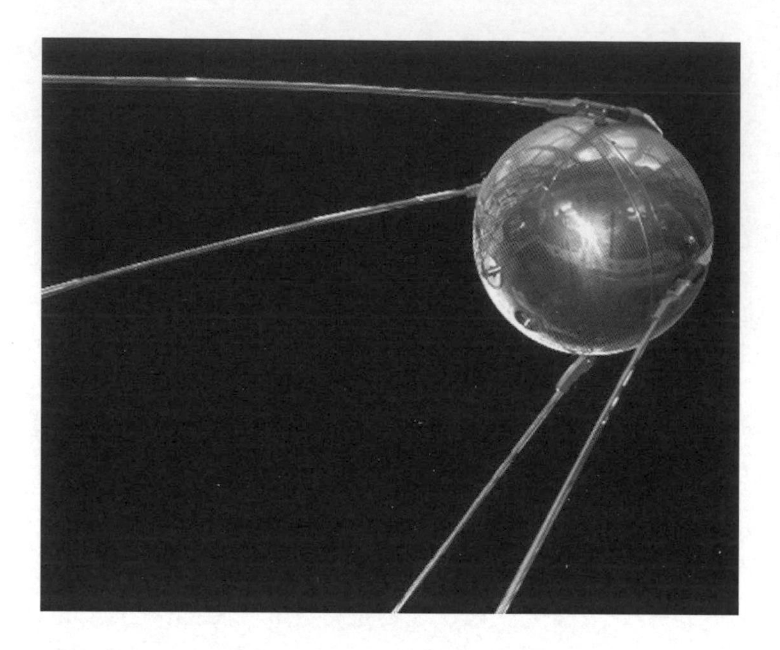

(*a*) (i) Sputnik 1 orbited Earth in 100 minutes.

Calculate the distance it travelled in this time. **2**

(ii) Although Sputnik 1 travelled at a constant speed in a circular orbit, it accelerated continuously.

Explain this statement. **2**

(*b*) Sputnik 1 transmitted radio signals a distance of 800 km to the surface of the Earth.

Calculate the time taken for the signals to reach the Earth's surface. **2**

21. (continued) *Marks*

(*c*) The graph shows how gravitational field strength varies with height above the surface of the Earth.

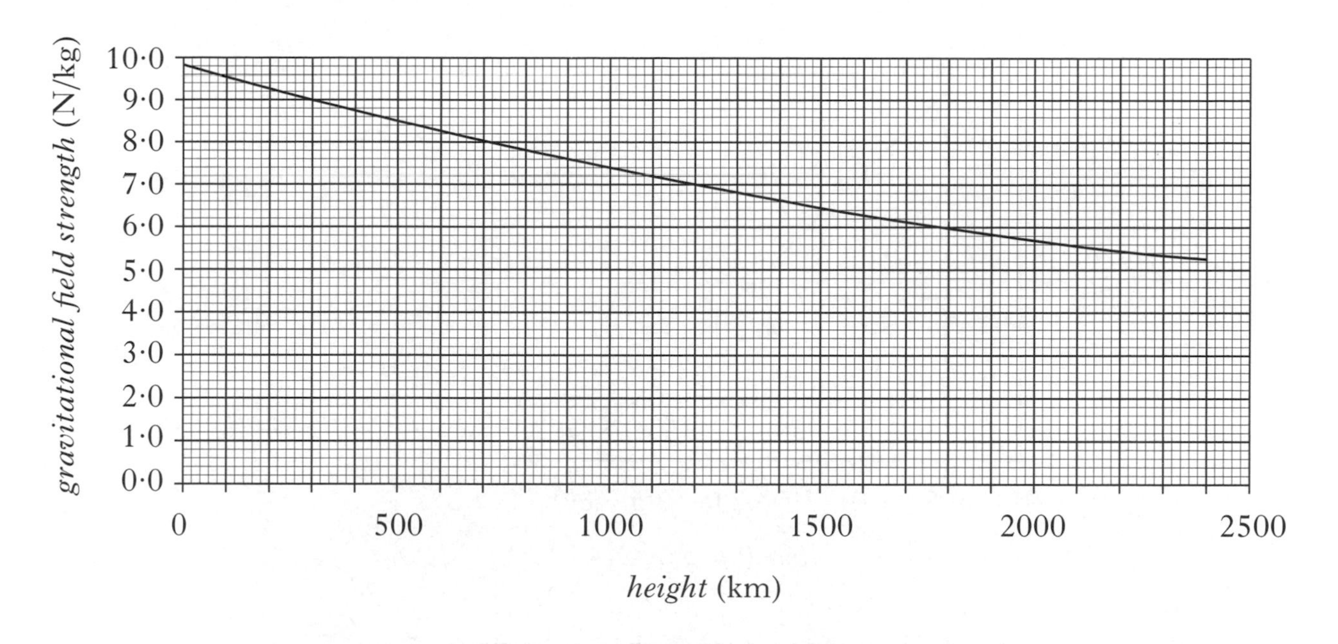

(i) Define the term **gravitational field strength**. **1**

(ii) What is the value of the gravitational field strength at a height of 800 km? **1**

(iii) Calculate the weight of Sputnik 1 at this height. **2**

(10)

[Turn over

Marks

22. A car of mass 700 kg travels along a motorway at a constant speed. The driver sees a traffic hold-up ahead and performs an emergency stop. A graph of the car's motion is shown, from the moment the driver sees the hold-up.

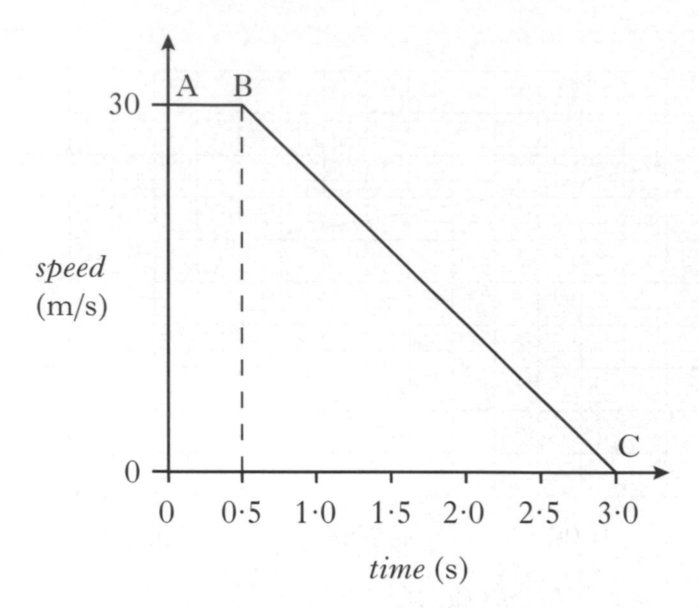

(*a*) Describe **and** explain the motion of the car between A and B. **2**

(*b*) Calculate the kinetic energy of the car at A. **2**

(*c*) State the work done in bringing the car to a halt between B and C. **1**

(*d*) Show by calculation that the magnitude of the unbalanced force required to bring the car to a halt between B and C is 8400 N. **2**

(7)

Marks

23. A student reproduces Galilleo's famous experiment by dropping a solid copper ball of mass 0·50 kg from a balcony on the Leaning Tower of Pisa.

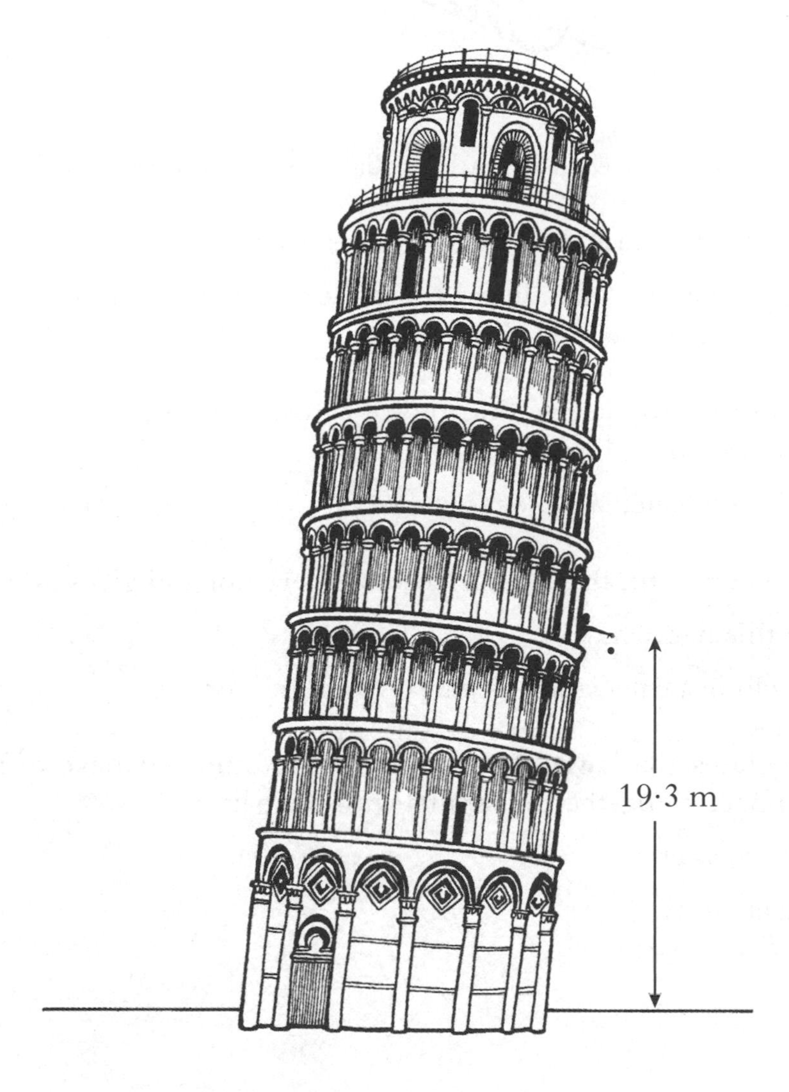

(*a*) (i) The ball is released from a height of 19·3 m.

Calculate the gravitational potential energy lost by the ball. **2**

(ii) Assuming that all of this gravitational potential energy is converted into heat energy **in the ball**, calculate the increase in the temperature of the ball on impact with the ground. **2**

(iii) Is the actual temperature change of the ball greater than, the same as or less than the value calculated in part (*a*)(ii)?

You **must** explain your answer. **2**

(*b*) The ball was made by melting 0·50 kg of copper at its melting point. Calculate the amount of heat energy required for this. **3**

(9)

Marks

24. A resistor is labelled: "10 Ω ± 10%, 3 W".

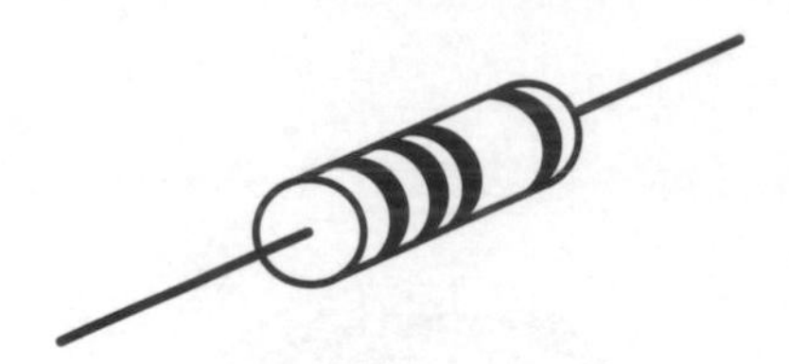

This means that the resistance value could actually be between 9 Ω and 11 Ω.

(*a*) A student decides to check the value of the resistance.

Draw a circuit diagram, including a 6 V battery, a voltmeter and an ammeter, for a circuit that could be used to determine the resistance. **3**

(*b*) Readings from the circuit give the voltage across the resistor as 5·7 V and the current in the resistor as 0·60 A.

Use these values to calculate the resistance. **2**

(*c*) During this experiment, the resistor becomes very hot and gives off smoke.

Explain why this happens.

You **must** include a calculation as part of your answer. **3**

(*d*) The student states that **two** of these resistors would not have overheated if they were connected together in parallel with the battery.

Is the student correct?

Explain your answer. **2**

(10)

Marks

25. The circuit shown switches a warning lamp on or off depending on the temperature.

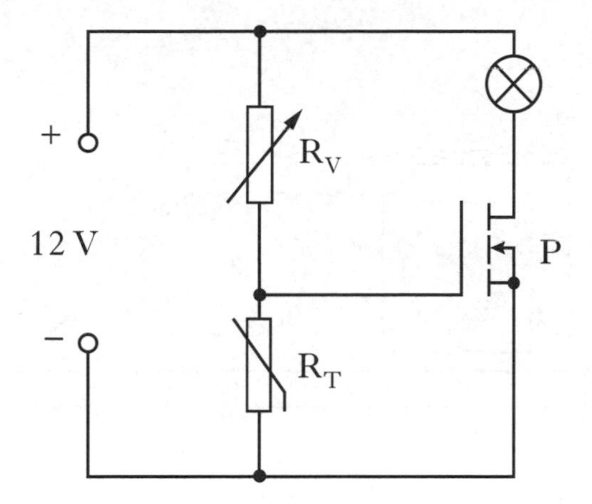

(*a*) Name component P. **1**

(*b*) As the temperature increases the resistance of thermistor R_T decreases. What happens to the voltage across R_T as the temperature increases? **1**

(*c*) When the voltage applied to component P is equal to or greater than 2·4 V, component P switches on and the warning lamp lights.

R_V is adjusted until its resistance is 5600 Ω and the warning lamp now lights.

At this point calculate:

(i) the voltage across R_V; **1**

(ii) the resistance of R_T. **2**

(*d*) The temperature of R_T now decreases.

Will the lamp stay on or go off?

You **must** explain your answer. **2**

(7)

[Turn over

Marks

26. An apparatus used to measure the speed of sound consists of a bright LED which flashes every 0·5 s and a loudspeaker which beeps at **exactly the same time** as the LED flashes.

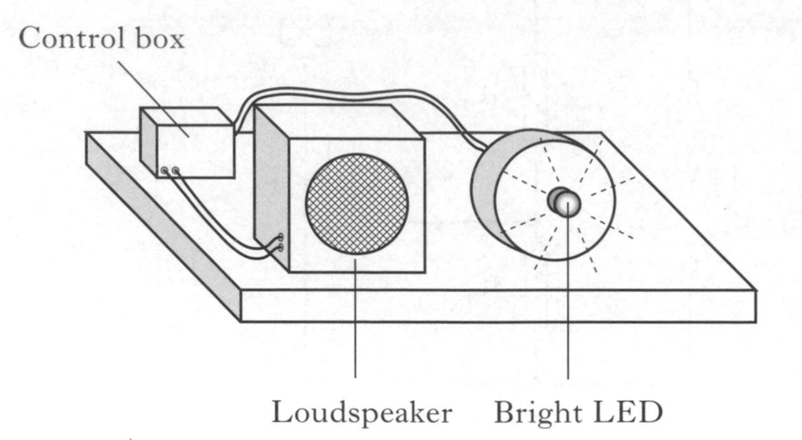

(*a*) A student standing beside the apparatus observes the beeps and flashes happening at exactly the same time.

Another student standing 88 m away does **not** observe them happening at the same time.

(i) Explain this observation. **1**

(ii) A third student 176 m away observes the beeps and flashes happening at exactly the same time.

Use this information to calculate a value for the speed of sound. **2**

(*b*) The circuit used to operate the LED and loudspeaker uses electric switches called relays.

Simplified diagrams of a relay are shown.

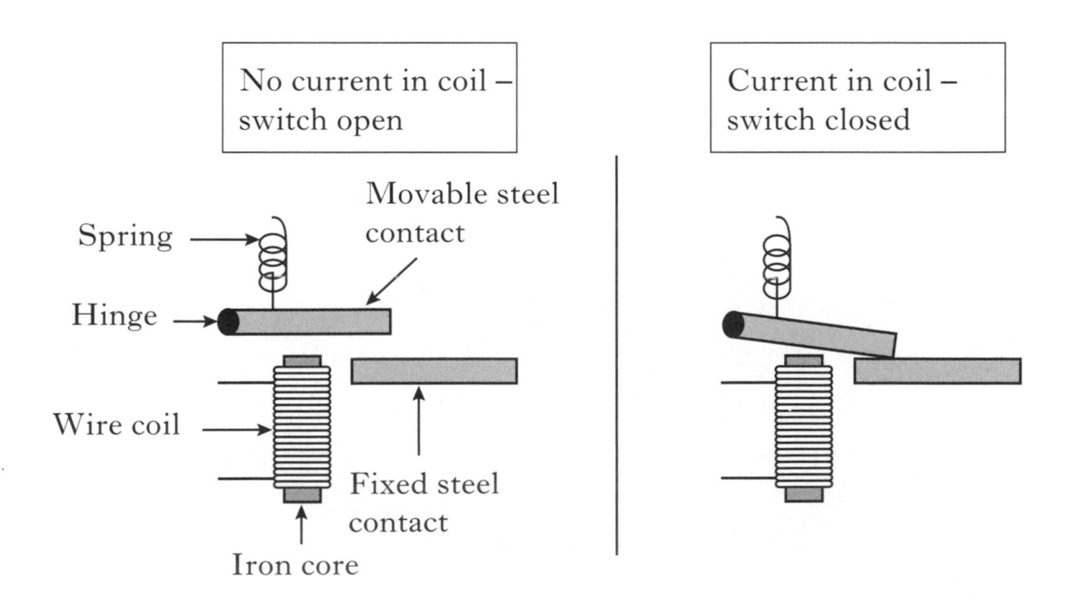

Explain why the steel contact moves when there is a current in the coil. **2**

26. (continued) *Marks*

(*c*) Part of the circuit for the apparatus is shown.

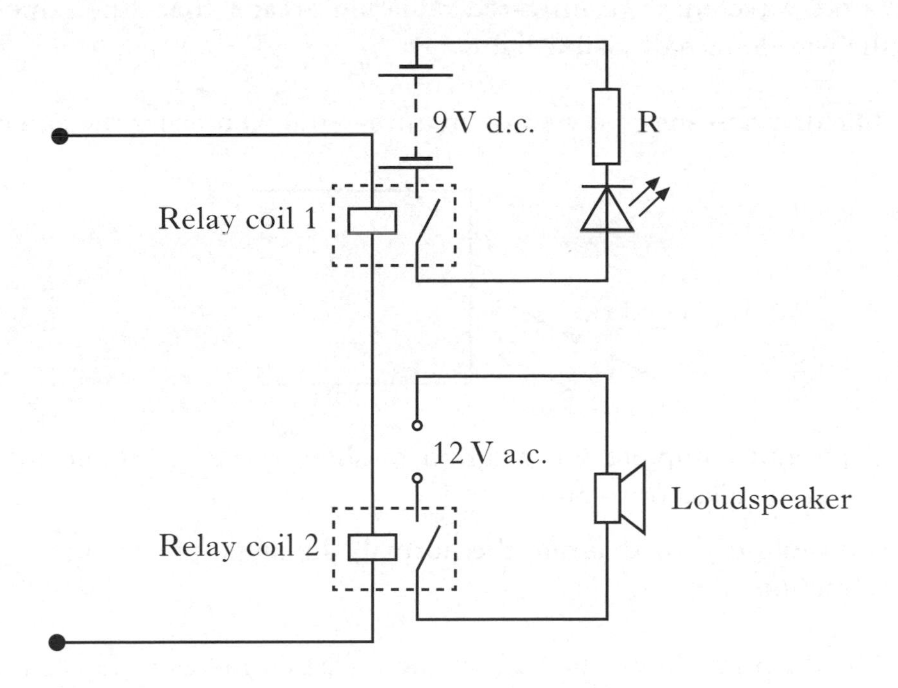

The LED is rated 6·0 V, 800 mA.

Calculate the value of resistor R. **3**

(*d*) The 12 V a.c. supply has a frequency of 850 Hz.

Using the value of the speed of sound from the data sheet, calculate the wavelength of the sound in air produced by the loudspeaker. **2**

(10)

[Turn over

Marks

27. Optical fibres are used to carry internet data using infra-red radiation.

(*a*) Is the wavelength of infra-red radiation greater than, the same as, or less than the wavelength of visible light? **1**

(*b*) The diagram shows a view of an infra-red ray entering the end of a fibre.

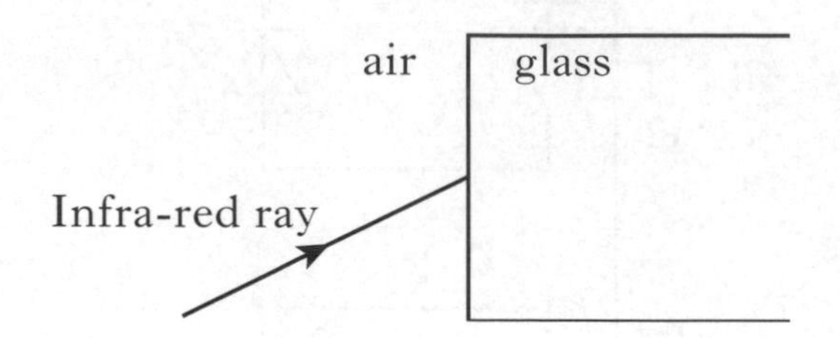

Copy and complete the diagram to show the path of the infra-red ray as it enters the glass from air.

Indicate on your diagram the normal, the angle of incidence and the angle of refraction. **2**

(*c*) The diagram shows the path of the infra-red ray as it passes through a section of the fibre.

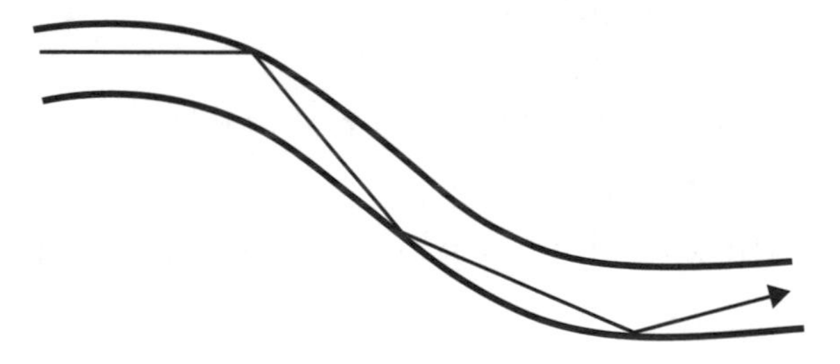

Name the effect shown. **1**

(4)

[Turn over for Question 28 on *Page twenty-two*

Marks

28. The picture shows a spy using a long range microphone and curved reflector to listen to conversations from some distance away.

(*a*) Copy and complete the following diagram to show the path of the sound waves. **2**

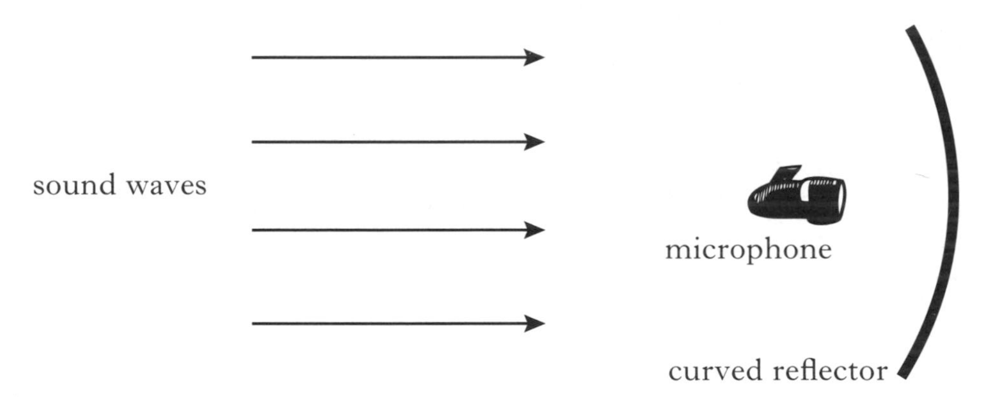

(*b*) Explain why using the curved reflector makes the sound detected by the microphone louder. **1**

(*c*) The microphone produces a signal of 24 mV which is applied to the input of an amplifier.

The output from the amplifier is 3·0 V.

Calculate the voltage gain of the amplifier. **2**

Marks

28. (continued)

(*d*) The spy needs spectacles to see distant objects clearly.

(i) What is the name given to this eye defect? **1**

(ii) What type of lens is needed to correct this defect? **1**

(*e*) The spy uses a magnifying lens of power +10 D to examine some photographs.

Calculate the focal length of this lens. **2**

(9)

[Turn over

Marks

29. A technician checks the count rate of a radioactive source. A graph of count rate against time for the source is shown. The count rate has been corrected for background radiation.

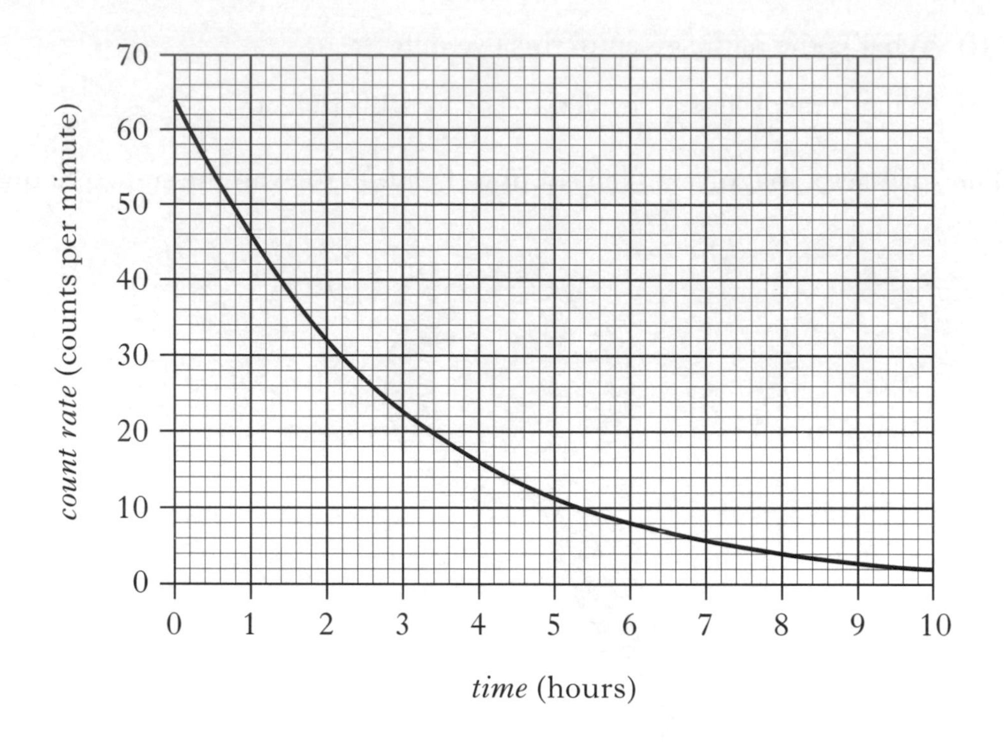

(*a*) Use the graph to determine the half-life of the source. **2**

(*b*) State **two** factors which can affect the background radiation level. **2**

(*c*) The source emits gamma rays. State what is meant by a gamma ray. **1**

(5)

Marks

30. An ageing nuclear power station is being dismantled.

© PA

(*a*) During the dismantling process a worker comes into contact with an object that emits 24 000 alpha particles in five minutes. The worker's hand has a mass of 0·50 kg and absorbs 6·0 μJ of energy.

Calculate:

(i) the absorbed dose received by the worker's hand; **2**

(ii) the equivalent dose received by the worker's hand; **2**

(iii) the activity of the object. **2**

(*b*) In a nuclear reactor, state the purpose of:

(i) the moderator; **1**

(ii) the containment vessel. **1**

(*c*) What type of nuclear reaction takes place in a nuclear power station's reactor? **1**

(9)

[*END OF QUESTION PAPER*]

Acknowledgements

Permission has been sought from all relevant copyright holders and Bright Red Publishing is grateful for the use of the following:
Picture of an exercise bike taken from www.multisportfitness.com © MultiSports, Inc (2008 page 11).

INTERMEDIATE 2 | ANSWER SECTION

SQA INTERMEDIATE 2 PHYSICS 2008–2012

PHYSICS INTERMEDIATE 2 2008

SECTION A

1. E
2. C
3. C
4. B
5. D
6. B
7. A
8. D
9. C
10. D
11. D
12. E
13. C
14. B
15. C
16. A
17. B
18. A
19. E
20. A

SECTION B

21. (*a*) $a = \frac{v - u}{t}$

$a = \frac{9}{2}$

$a = 4{\cdot}5\ m/s^2$

(*b*) $F = m \times a$
$F = 15 \times 4{\cdot}5$
$F = 67{\cdot}5\ N$

(*c*) d = area under graph
$d = (0{\cdot}5 \times 9 \times 2) + (10 \times 9) + (0{\cdot}5 \times 9 \times 1)$
$d = 9 + 90 + 4{\cdot}5$
$d = 103{\cdot}5\ m$

(*d*) $P = \frac{1}{f}$

$P = \frac{1}{0{\cdot}2}$

$P = 5\ D$

22. (*a*)

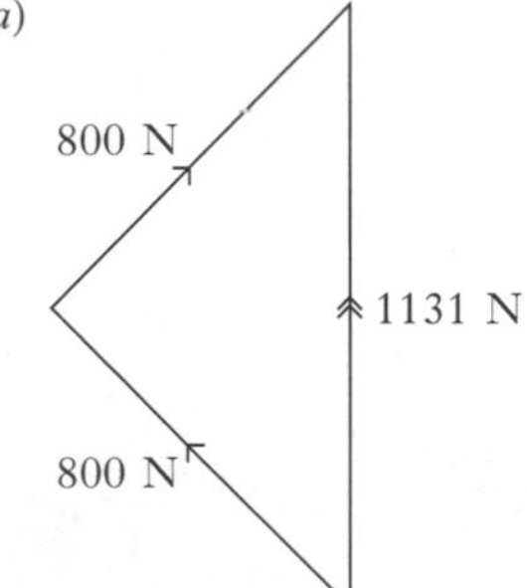

or by calculation

$d = \sqrt{800^2 + 800^2}$
$= 1131\ N$

(*b*) (i) $W = mg$
$= 180 \times 10$
$= 1800\ N$

(ii) resultant $= 2700 - 1800 = 900\ N$

$= \frac{F}{m}$

$= \frac{900}{180}$

$a = 5\ m/s^2$

23. (*a*) (i) $E_w = F \times d$
$E_w = 300 \times 1{\cdot}5$
$E_w = 450\ J$

(ii) $E = 450 \times 500 = 225000\ J$

$P = \frac{E}{t}$

$P = \frac{225000}{5 \times 60}$

$P = 750\ W$

(*b*) (i) $E = c\ m\ \Delta T$
$450 \times 500 = 902 \times 12 \times \Delta T$
$\Delta T = 21\ °C$

(ii) energy is lost to the surrounding air

24. (*a*) $E_p = mgh$
$E_p = 750 \times 10 \times 7{\cdot}2$
$E_p = 54000\ J$

(*b*) (i) 54000 J

(ii) $E_K = \frac{1}{2}mv^2$

$54000 = 0{\cdot}5 \times 750 \times v^2$
$v = 12\ m/s$

25. (*a*) $P = I^2 R$
$2 = I^2 \times 50$
$I^2 = 0{\cdot}04$
$I = 0{\cdot}2\ A$

(*b*) (i) $\frac{1}{R_t} = \frac{1}{R_1} + \frac{1}{R_2}$

$\frac{1}{R_t} = \frac{1}{60} + \frac{1}{30}$

$R_t = 20\ \Omega$

(ii) $P = \frac{V^2}{R}$

$P = \frac{9^2}{60}$

$= 1{\cdot}35\ W$

$P = \frac{V^2}{R}$

$P = \frac{9^2}{30}$

$= 2{\cdot}7\ W$

(iii) 30 Ω resistor will overheat

(*c*) none

26. (*a*) Sound energy to electrical energy

(*b*) (i) None
(ii) Greater

(*c*) $v = f\lambda$
$340 = 850 \times f\lambda$
$\lambda = 0{\cdot}4\ m$

(*d*) (i) If light inside the prism strikes the surface at an angle greater than the critical angle it will be totally internally reflected.

(ii)

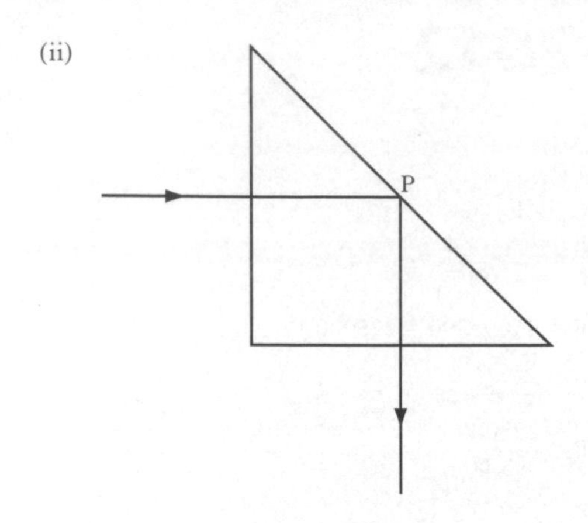

27. (*a*) (i) The resistance of LDR drops (with light level rise)
V across R rises until <u>MOSFET</u> switches <u>on the motor</u>

(ii) to set the light level at which the blind closes.

(*b*) (i) 3000 Ω

(ii) $V_1 = \left(\frac{R_1}{R_1 + R_2}\right)V_S$

$V = \left(\frac{600}{600 + 3000}\right) \times 12$

$V = 2$ V

(iii) Since V < 2·4 V transistor will not switch on so blinds do not shut.

28. (*a*) (i) to limit current in/voltage across the LED

(ii) Vr = 12 – 2 = 10 V

$R = \frac{V}{I}$

$R = \frac{10}{0{\cdot}02}$

R = 500 Ω

(iii) I = 10 × 20
= 200 mA
= 0·2 A

(*b*) $\frac{n_s}{n_p} = \frac{V_s}{V_p}$

$\frac{n_s}{200} = \frac{84}{12}$

$n_s = 1400$ (turns)

29. (*a*) Converging/convex

(*b*)

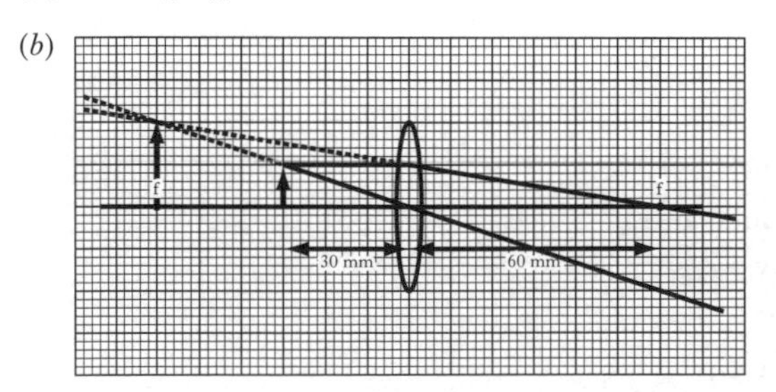

(*c*) Make thinner/or less curved

(*d*) Long sight

30. (*a*) Count rate increases
Air is more easily penetrated/less metal to be penetrated

(*b*) Gamma
penetrates best/other two would not penetrate steel

(*c*) x-rays longer/gamma shorter

31. (*a*) time taken for half of the radioactive atoms to decay or activity to decrease by half

(*b*)

Days	activity
0	64
2·7	32
5·4	16
8·1	8
10.8	4
13·5	**2 kBq**

(*c*) *Any two from:*
- shielding
- limiting time of exposure
- increasing distance

(*d*) (i) $H = w_r D$
= 20 × 10 mGy
= 200 mSv

(ii) Tissue type

PHYSICS INTERMEDIATE 2 2009

SECTION A

1.	C	11.	B
2.	C	12.	D
3.	E	13.	A
4.	A	14.	D
5.	B	15.	C
6.	B	16.	E
7.	D	17.	E
8.	E	18.	A
9.	B	19.	D
10.	E	20.	C

SECTION B

21. (*a*) $E_p = mgh$
$= 2000 \times 10 \times 540$
$= 10800000$ J ($1{\cdot}08 \times 10^7$ J)

(*b*) $E_k = \frac{1}{2}mv^2$
$64000 = 0{\cdot}5 \times 2000 \times v^2$
$v^2 = 64$
$v = 8$ m/s

(*c*) (i) $P = IV$
$45600 = I \times 380$
$I = 120$ A

(ii) $E = Pt$
$= 45600 \times 60 \times 60$
$= 1{\cdot}64 \times 10^8$ J

22. (*a*) $a = \frac{(v-u)}{t}$

$a = \frac{(3-0)}{5}$

$a = 0{\cdot}6$ m/s^2

(*b*) $F = ma$
$F = 40 \times 0{\cdot}6$
$= 24$ N

(*c*) There is an unbalanced force/friction, which acts against the motion.

23. (*a*) width/length of card (d)
time taken for card to cut beam (t)

$v = \frac{d}{t}$

or

$v = \frac{\text{length of card}}{\text{time taken for card to cut beam}}$

(*b*) (i) $p = mv$
$= (5 \times 10^{-4} + 0{\cdot}3) \times 0{\cdot}35$
$= 0{\cdot}105$ kg m/s

(ii) $v = \frac{p}{m} = \frac{0{\cdot}105}{5 \times 10^{-4}}$
$= 210$ m/s

(*c*) (i) $a = \frac{(v-u)}{t}$

$10 = \frac{(v-0)}{0{\cdot}2}$

$v = 2$ m/s

(ii) $d = \bar{v}t$
$= 1 \times 0{\cdot}2$
$= 0{\cdot}2$ m

24. (*a*) $E_h = cm\Delta T$
$= 4180 \times 0{\cdot}1 \times 15$
$= 6270$ J

(*b*) $E_h = ml$
$= 0{\cdot}1 \times 3{\cdot}34 \times 10^5$
$= 3{\cdot}34 \times 10^4$ J

(*c*) (i) $33400 + 6270 = 39670$ J

$E = Pt$
$39670 = 125 \times t$
$t = 317{\cdot}(36)$ s

(ii) Heat energy will be gained from surroundings/other food etc.
Therefore more energy must be removed.

25. (*a*) (i) $I = 0{\cdot}075$ A
$V = IR$
$4{\cdot}2 = 0{\cdot}075 \times R$
$R = 56\ \Omega$

(ii) stays the same

$\frac{1{\cdot}3}{0{\cdot}023} = 56{\cdot}5$ $\quad$ $\frac{3{\cdot}6}{0{\cdot}064} = 56{\cdot}25$

or as the voltage increases the current increases by the same ratio
or because it's a straight line through the origin

(*b*) (i) $R_t = R_1 + R_2$
$= 270 + 390$
$= 660\ \Omega$

(ii) $\frac{1}{R_t} = \frac{1}{R_1} + \frac{1}{R_2}$
$= \frac{1}{33} + \frac{1}{56}$
$= 0{\cdot}048$

$R_t = 20{\cdot}76\ \Omega$

26. (*a*) $\frac{I_p}{I_s} = \frac{V_s}{V_p}$

$\frac{I_p}{I} = \frac{5}{230}$

$I_p = 0{\cdot}022$ A

(*b*) (i) $P = \frac{V^2}{R}$

$10 = \frac{9^2}{R}$

$R = 8{\cdot}1\ \Omega$

(ii) $V_g = \frac{V_o}{V_i}$

$= \frac{9}{1{\cdot}5}$

$= 6$

(*c*) $Eff\% = \frac{P_o}{P_i} \times 100$

$= \frac{20}{25} \times 100$

$= 80\%$

27. (*a*) (i) short sight = the image is in focus in front of the retina
or
cannot see distant objects clearly

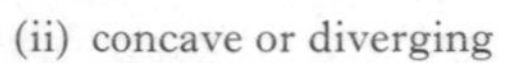

(ii) concave or diverging

(iii) $P = \frac{1}{f}$

$= (-)\frac{1}{0{\cdot}18}$

$= (-)\ 5{\cdot}6\ \text{D}$

(*b*) (i) refraction = the change in the speed or wavelength of light as it passes between two media (of different densities)

(ii) $v = f\lambda$

$3 \times 10^8 = f \times 7 \times 10^{-7}$

$f = 4{\cdot}29 \times 10^{14}$ Hz

$f = 4 \times 10^{14}$ Hz

(*c*) (i)

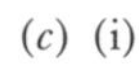

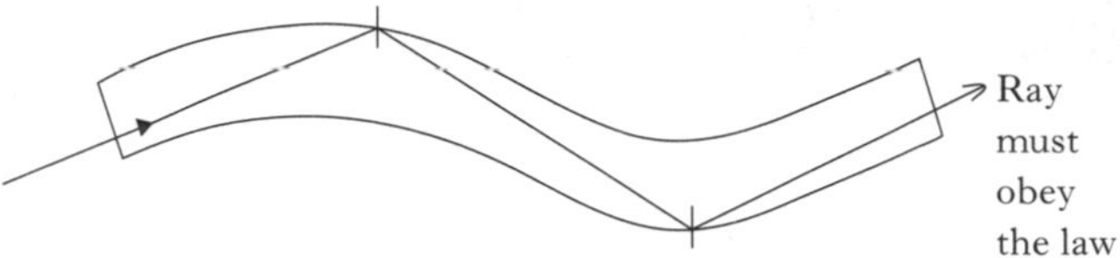

Ray must obey the law of reflection

Appropriate number of reflections

(ii) (total internal) reflection (TIR)

28. (*a*) $v = \frac{d}{t}$

$340 = \frac{d}{2\times10^{-3}}$

$d = 0{\cdot}68$ m

$\therefore d = 0{\cdot}34$ m one way

(*b*) $(f = \frac{1}{T})$

$f = \frac{1}{0{\cdot}125}$

$f = 8$ Hz

(*c*) I = 200 mA

$P = IV$

$= 200 \times 10^{-3} \times 12$

$= 2{\cdot}4$ W

(*d*) (i) (the resistor) stops too large a current (flowing through the LED) or too large a voltage across the LED

(ii) $V = 12 - 3{\cdot}5 = 8{\cdot}5$ (V)

$V = IR$

$8{\cdot}5 = 200 \times 10^{-3} \times R$

$R = 42{\cdot}5\ \Omega$

29. (*a*) (i) equipment:

source, paper and suitable radiation detector and counter

measurements:

measure the count rate from each of the sources with paper and without paper between the source and the detector

explanation:

the source which produced a decreased count rate with paper is the alpha source

(ii) Cover the front window with a few mm of aluminium to stop beta.

(*b*)

Time	Activity
0	1
28	$\frac{1}{2}$
56	$\frac{1}{4}$
84	$\frac{1}{8}$
112	$\frac{1}{16}$

112 years

(*c*) (i) $H = Dw_R$

$= 20 \times 10^{-6} \times 20$

$= 400\ \mu S_v$

or $400 \times 10^{-6}\ S_v$

(ii) increase distance (eg use tongs)

shielding (lead apron/gloves)

PHYSICS INTERMEDIATE 2 2010

SECTION A

1.	E	11.	B
2.	D	12.	B
3.	B	13.	C
4.	D	14.	A
5.	D	15.	A
6.	C	16.	D
7.	E	17.	C
8.	C	18.	E
9.	D	19.	A
10.	D	20.	E

SECTION B

21. (*a*) $a = \frac{v - u}{t}$

$= \frac{6 - 0}{60}$

$= 0{\cdot}1\ m/s^2$

(*b*) s = area under graph

$= (0{\cdot}5 \times 60 \times 6) + (40 \times 6)$

= 420 m

(*c*) $v = \frac{s}{t}$

$= \frac{420}{100}$

= 4·2m/s

(*d*) W = mg

$= 400 \times 10$

= 4000 N

(*e*) F = ma

$= 400 \times 0{\cdot}1$

= 40 (N)

Upward force = 4000 + 40

= 4040 N

22. (*a*) p before = p after

$(2{\cdot}0 \times 10^{-3} \times 4) = 3{\cdot}2 \times 10^{-3}\ v$

v = 2·5 m/s

(*b*)

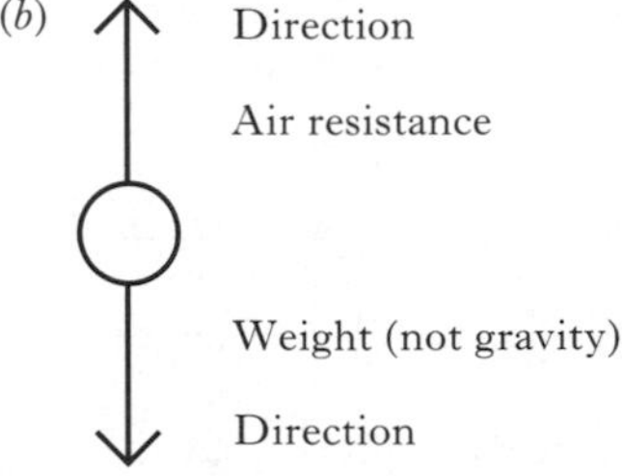

(*c*) Q = It

$I = \frac{1650}{0{\cdot}15}$

$= 1{\cdot}1 \times 10^4\ A$

(*d*) Light travels faster than sound

23. (*a*) $E_h = cm\Delta T$

$c = \frac{2{\cdot}59 \times 10^7}{60 \times [(307 - (-173)]}$

= 899 J/kg°C

(*b*) $P = \frac{E}{t}$

$t = \frac{2{\cdot}59 \times 10^7}{1440}$

= 18000 s

(*c*) $\frac{288000}{1440}$

= 200 (rocks)

(*d*) It would be easier

Gravitational field strength at the surface of Mercury is less than that at the surface of Earth

or

Weight of rocks on Mercury is smaller than their weight on Earth

or

Gravity is less on Mercury

24. (*a*) $R_T = R_1 + R_2 = 8 + 24 = 32\ \Omega$

V = IR

$I = \frac{6}{32}$

I = 0·19 A

(*b*) $V_2 = \left(\frac{R_2}{R_1 + R_2}\right) V_S$

$V_2 = \left(\frac{8}{8 + 24}\right) 6$

$V_2 = 1{\cdot}5\ V$

or

V = IR

$= 0{\cdot}19 \times 8$

= 1·5 V

(*c*) Voltage across 8 Ω resistor would decrease

The 8 Ω resistor now has a smaller proportion of the total resistance

or less current in the circuit

25. (*a*) a.c. (source)

changing magnetic field **or** changing current is necessary (to induce voltage)

(*b*) P = IV

$= 0{\cdot}5 \times 12$

= 6W

(*c*) P = IV

$= 0{\cdot}23 \times 23$

= 5·3W

(*d*) percentage efficiency $= \frac{\text{useful } P_o}{Pi} \times 100$

$= \frac{5{\cdot}3}{6} \times 100$

= 88(%)

(*e*) $\frac{N_S}{N_P} = \frac{V_S}{V_P}$

$V_S = \frac{3000 \times 12}{1500}$

= 24V

26. (*a*) Y (n-channel enhancement) MOSFET

Z Lamp

(*b*) (Resistance) decreases

(*c*) (As resistance of thermistor decreases) voltage across thermistor decreases.

V across X increases

When it reaches 1·8V MOSFET $V_{(transistor)}$ switches on

(Bulb lights and) buzzer sounds

(*d*) To allow switch on temperatures to be varied

27. (*a*) s = vt

$t = \frac{51}{340}$

= 0·15 s

(*b*) (i) Longitudinal

(ii) A transverse wave is one in which the particles vibrate at right angles to the direction of the wave.
A longitudinal wave is one in which the particles vibrate parallel to the direction of the wave.

(iii) Energy

(*c*) (i) $P = \frac{V^2}{R}$

$R = \frac{315^2}{2400}$

$= 41 \cdot 34\ \Omega$

(ii) *Any two from:*

- independent switching or one off/others stay on
- to ensure that 315 V is across each bulb
- if they were in series the necessary voltage would be too high

28. (*a*) (i) $v = f\lambda$

$f = \frac{3 \times 10^8}{0 \cdot 06}$

$= 5 \times 10^9$ Hz

(ii) $T = \frac{1}{f}$

$T = \frac{1}{5 \times 10^9}$

$= 2 \times 10^{-8}$ s

(*b*) Signals received at same time
Radio waves and microwaves have same speed

(*c*)

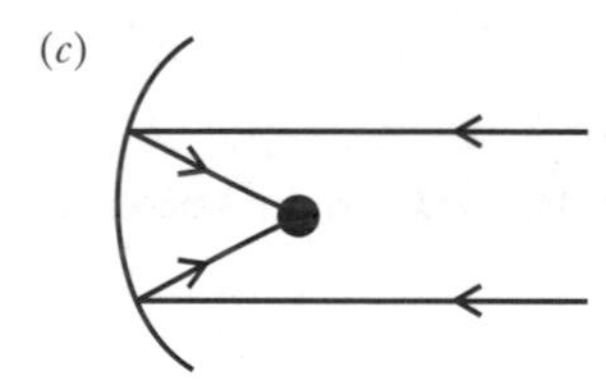

Diagram must have minimum of two rays. All rays drawn must come to a distinct focus or incoming waves are focussed towards one point.

Curved reflector gives point where energy is maximised

29. (*a*) A particle containing two protons and two neutrons
or
A helium nucleus

(*b*) The gain/loss of electrons by an atom

(*c*) $4800 \xrightarrow{1} 2400 \xrightarrow{2} 1200 \xrightarrow{3} 600 \xrightarrow{4} 300$ or equivalent

$4 \times 2 \cdot 5 = 10$ hours

(*d*) $A = \frac{N}{t}$

$N = 1200 \times 2 \times 60$

$= 144\,000$ (decays)

(*e*) Source may also emit β and/or γ radiation

30. (*a*) (i) Slows neutrons

(ii) Absorbs neutrons

(*b*) $P = \frac{E}{t}$

$E = 1 \cdot 4 \times 10^9 \times 60 \times 60$

$= 5 \cdot 0 \times 10^{12}$ (J)

Number of fissions $= \frac{5 \cdot 0 \times 10^{12}}{2 \cdot 9 \times 10^{-11}}$

$= 1 \cdot 7 \times 10^{23}$

(*c*) Any valid advantage eg much greater energy per kg of fuel compared to other sources
or
No greenhouse gases emitted or equivalent

PHYSICS INTERMEDIATE 2 2011

SECTION A

1.	D	11.	D
2.	B	12.	B
3.	E	13.	B
4.	D	14.	D
5.	E	15.	E
6.	B	16.	C
7.	A	17.	C
8.	E	18.	D
9.	B	19.	A
10.	A	20.	A

SECTION B

21. (*a*) $s = vt$

$\therefore\ t = \frac{11}{20}$

$= 0 \cdot 55$ s

(*b*) $a = \frac{v - u}{t}$

$\therefore\ v = 10 \times 0 \cdot 55$

$= 5 \cdot 5$ m/s

(*c*)

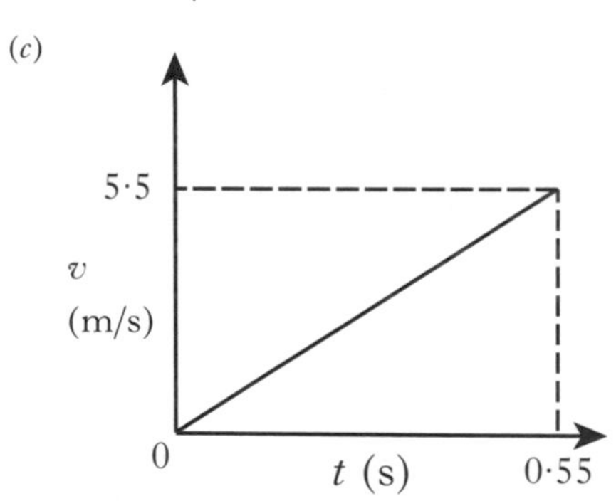

(*d*) s = area under graph **or** $s = \bar{v}t$

$s = \frac{1}{2} \times 0 \cdot 55 \times 5 \cdot 5$ $s = \left(\frac{5 \cdot 5}{2}\right) \times 0 \cdot 55$

$s = 1 \cdot 5$ m $s = 1 \cdot 5$ m

22. (*a*) (i) Acceleration is the change of velocity (not speed) in unit time

(ii) Direction of satellite is (continually) changing
or
Velocity of satellite is (continually) changing
or
There is an unbalanced (not 'resultant') force on the satellite

(*b*) $F = 12 - 2 = 10$N

$F = ma$

$\therefore 10 = 50a$

$a = 0 \cdot 2$ m/s^2

Direction is right

23. (*a*) W = mg

= 50,000 × 10

= 500,000N

(*b*) 500,000N

(*c*) For scale drawing accept

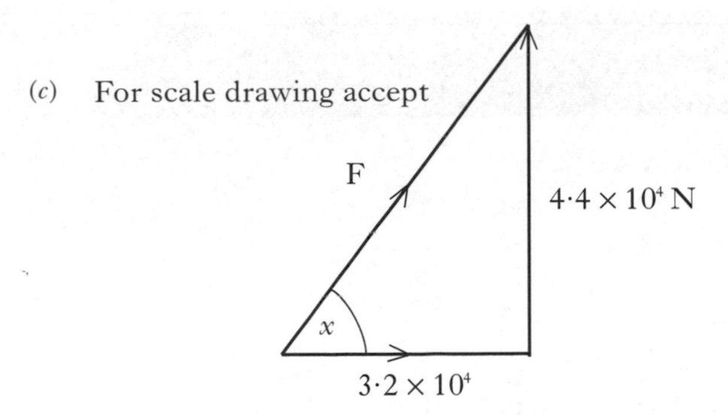

$F^2 = (3{\cdot}2 \times 10^4)^2 + (4{\cdot}4 \times 10^4)^2$
$F = 5{\cdot}4 \times 10^4$N
$\tan x = \frac{4{\cdot}4 \times 10^4}{3{\cdot}2 \times 10^4}$
$x = 54°$
$F = 5{\cdot}4 \times 10^4$ N at 036°

(*d*) $H = DW_R$
$= 15 \times 10^{-6} \times 1$
$= 1{\cdot}5 \times 10^{-5}$ Sv (15×10^{-6})

(*e*) Ionisation is when an atom gains or loses electrons
must have — one only needed

24. (*a*) (i) $(33 - 21) = 12°C$

(ii) (120,000 – 12,000) = 108,000 J

(iii) $E_h = cm\Delta T$
$108{,}000 = c \times 2{\cdot}0 \times 12$
$c = 4{,}500$ J/kg°C

(*b*) (i) Heat lost to surroundings (or similar)
or water not evenly heated (or similar)
Measured value of E_h too large **or** ΔT too small

(ii) Insulate beaker
or Put lid on beaker
or Stir water
or Fully immerse heater

(*c*) $E = Pt$
$108{,}000 = P \times 5 \times 60$
$P = 360$ W

25. (*a*) $\frac{1}{R_T} = \frac{1}{R_1} + \frac{1}{R_2}$
$= \frac{1}{4} + \frac{1}{2}$
$\therefore R_T = 1{\cdot}3\,\Omega$

(*b*) $R_T = R_1 + R_2$
$= 1{\cdot}3 + 6$
$= 7{\cdot}3\,\Omega$

(*c*) (Voltage across $2\,\Omega$ resistor = Voltage across $4\,\Omega$ resistor)
$V = IR$
$= 0{\cdot}1 \times 4$ (**or** $0{\cdot}2 \times 2$)
$= 0{\cdot}4$V

26. (*a*) dc – electron flows around a circuit in one direction only
ac – electrons' direction changes/reverses continuously

(*b*) (i) ac **or** mains **or** one on left

(ii) Transformers are used to change the magnitude of an (alternating) voltage **or** current

(iii) Percentage efficiency = $\frac{useful\ P_o}{P_i} \times 100$

$useful\ P_o = \frac{30}{100} \times 50$
$= 15$ W

27. (*a*) To reduce current in LED
or
To reduce voltage across LED
or
To reduce power to LED

(*b*) $V = 6 - 2 = 4V$
$V = IR$
$\therefore R = \frac{4}{0{\cdot}1}$
$= 40\,\Omega$

(*c*) $P = I^2R$
$= (0{\cdot}1)^2 \times 40$
$= 0{\cdot}4$W

28. (*a*) (i) $P = IV$
$= 0{\cdot}4 \times 10^{-3} \times 0{\cdot}5$
$= 2 \times 10^{-4}$W

(ii) $= \frac{4 \times 10^{-3}}{2 \times 10^{-4}}$
= 20 (cells)

(*b*) Light → electric (al)

(*c*) $v = f\lambda$
$\therefore \lambda = \frac{v}{f}$
$= \frac{3 \times 10^8}{6{\cdot}7 \times 10^{14}}$
$= 4{\cdot}5 \times 10^{-7}$m

29. (*a*) (i) P – Ultraviolet **or** UV
Q – Microwaves

(ii) $s = vt$
$\therefore t = \frac{s}{v}$
$= \frac{4{\cdot}50 \times 10^{12}}{3 \times 10^8}$
$= 1{\cdot}5 \times 10^4$ s

(iii) Decreases

(*b*) $Q = It$
$\therefore I = \frac{Q}{t}$
$= \frac{360}{60}$
$= 6$A

30. (*a*) $P = \frac{1}{f}$
$= \frac{1}{0{\cdot}03}$
= 33 D

(*b*)

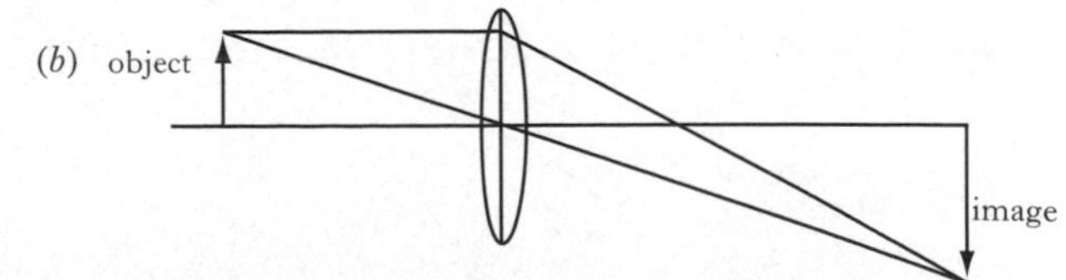

(*c*) Long sight
Converging lens brings light rays to focus on retina by reducing focal length (or equivalent).

31. (*a*) $N = At$

$= \underline{300 \times 10^{-6} \times 24 \times 60 \times 60}$

$= 26$ (decays)

(*b*) $2400 \rightarrow 1200 \rightarrow 600 \rightarrow 300$
$3 \times 5{,}730 = 17{,}190$ years

(*c*) An electron

(*d*) A helium nucleus **or** equivalent eg 2p + 2n

(*e*) Greater

(*f*) (i) (Aluminium) would stop α particles also

(ii) *Any two from:*
Shielding/Short times/Point away from people/ Increased distance/Wash hands

PHYSICS INTERMEDIATE 2 2012

SECTION A

1.	D	**11.**	D
2.	A	**12.**	A
3.	D	**13.**	B
4.	B	**14.**	D
5.	C	**15.**	D
6.	A	**16.**	E
7.	C	**17.**	C
8.	B	**18.**	C
9.	C	**19.**	E
10.	A	**20.**	E

21. (*a*) (i) $d = vt$
$8300 \times 100 \times 60$
$= 49\,800\,000$ m

(ii) (As orbit is circular) direction changes / **or** unbalanced force exists
so velocity changes.

(*b*) $d = vt$
$800 \times 1000 = 300\,000\,000\, t$
$t = 0.0027$ s

(*c*) (i) The weight of 1 kg **or** Weight per unit mass **or** Earth's pull per kg.

(ii) 7.8 N/kg

(iii) $W = mg$
$= 84 \times 7.8$
$= 660$ N

22. (*a*) Car continues at a constant speed during this time. AB represents driver's reaction time **or** the forces are balanced (or equivalent).

(*b*) $E = \frac{1}{2}mv^2$

$= 0.5 \times 700 \times 30^2$
$= 315\,000$ J

(*c*) 315 000 J

(*d*) $a = \frac{v - u}{t}$

$= (0 - 30)/2.5$
$(-)12(\text{m/s}^2)$

$F = ma$
$= 700 \times 12$
$= 8400$ N

or

$d =$ area under graph
$= 0.5 \times 2.5 \times 30$
$= 37.5$ (m)

$E_W = Fd$
$315\,000 = F \times 37.5$
$F = 8400$ N

23. (*a*) (i) $E_P = mgh$
$= 0.5 \times 10 \times 19.3$
$= 96.5$ J

(ii) $E_H = cm\Delta T$
$96.5 = 386 \times 0.50 \times \Delta T$
$\Delta T = 0.5°$ C

(iii) Less than.
Some heat is lost to surroundings/or equivalent.

(*b*) $E_h = ml$
$= 0.50 \times (2.05 \times 10^5)$
$= 102\,500$ J

24. (*a*)

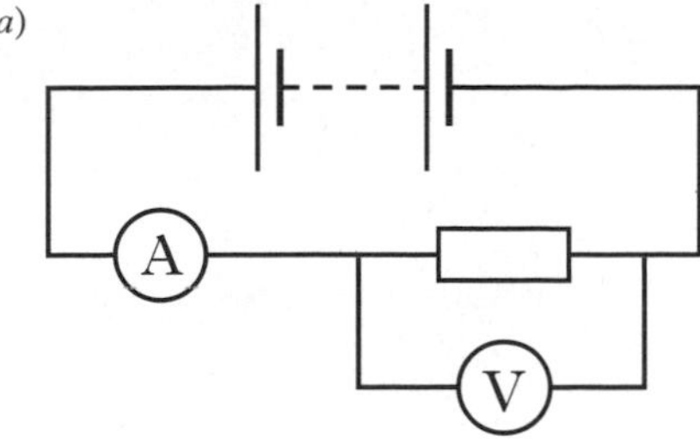

(*b*) $V = IR$
$5.7 = 0.60 \times R$
$R = 9.5\ \Omega$

(*c*) $P = VI$
$P = 5.7 \times 0.60$
$P = 3.42$ W
This is greater than the 3W or labelled power rating (so it overheats).

(*d*) No

In parallel the voltage is still the same/6V across each resistor

So power is the same

25. (*a*) MOSFET

(*b*) (Voltage) falls/decreases

(*c*) (i) $12 - 2.4 = 9.6$ V

(ii) $\frac{V_1}{V_2} = \frac{R_1}{R_2}$

$\frac{9.6}{2.4} = \frac{5600}{R_2}$

$R_2 = 1400\ \Omega$

(*d*) (Lamp) stays on

(Temperature falls)

R_T rises

V_T rises

$V_T > 2.4$ V **or** switching voltage

26. (*a*) (i) Speed of sound (much) less than speed of light (or similar)

(ii) $d = \bar{v}t$
$176 = \bar{v} \times 0.5$
$\bar{v} = 352$ m/s

(*b*) The current creates a magnetic field around the coil
The steel contact is attracted by the (magnetised) coil

(*c*) $V_R = 9 - 6 = 3$ V

$V = IR$
$3 = 800 \times 10^{-3} \times R$
$R = 3.75\ \Omega$

(*d*) $v = f\lambda$
$340 = 850 \times \lambda$
$\lambda = 0.4$ m

27. (*a*) Greater

(*b*)

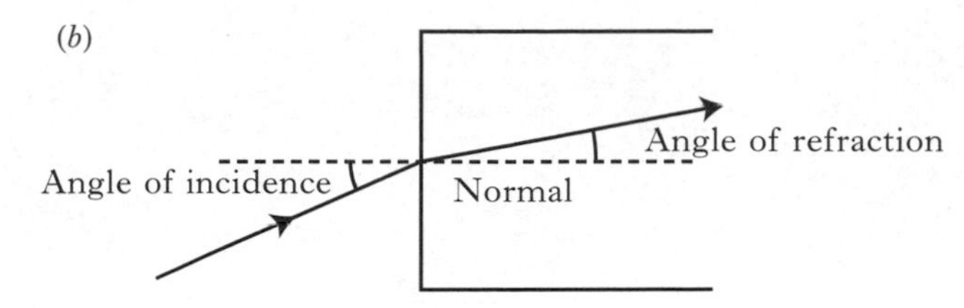

(*c*) Total internal reflection

28. (*a*)

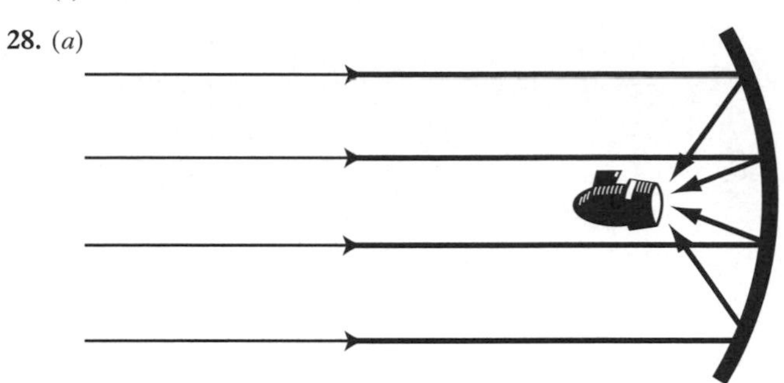

(*b*) More energy **or** power **or** amplitude is received (at the microphone.)

(*c*) $V_{gain} = \frac{V_o}{V_i}$

$V_{gain} = \frac{3}{0.024}$

$V_{gain} = 125$

(*d*) (i) Short sight

(ii) Diverging/concave lens

(*e*) $P = \frac{1}{f}$

$10 = \frac{1}{f}$

$f = 0.1$ m (10 cm)

29. (*a*) Half-life = 2 hours

(*b*) Any two valid answers.

(*c*) A type of <u>electromagnetic</u> radiation/wave/ray.

30. (*a*) (i) $D = E / m$
$= 0.000006/0.50$
$= 0.000012$ Gy

(ii) $H = Dw_R$
$= 0.000012 \times 20$
$= 0.00024$ Sv

(iii) $A = N / t$
$= 24{,}000/(5 \times 60)$
$= 80$ Bq

(*b*) (i) The moderator <u>slows neutrons</u>.

(ii) The containment vessel prevents/reduces release of radiations **or** radioactive gases **or** radioactive substances etc.

(*c*) Fission or Chain reaction.

BrightRED
PUBLISHING

Published by Bright Red Publishing Ltd, 6 Stafford Street, Edinburgh, EH3 7AU
Tel: 0131 220 5804, Fax: 0131 220 6710, enquiries: sales@brightredpublishing.co.uk,
www.brightredpublishing.co.uk

Official SQA answers to 978-1-84948-279-0
2008-2012